Eduardo Luersen, James Wilson (eds.)
Digital Culture & Society (DCS)

Digital Culture & Society | Volume 19

Editorial

Digital Culture & Society is a refereed, international journal, fostering discussion about the ways in which digital technologies, platforms and applications reconfigure daily lives and practices. It offers a forum for critical analysis and inquiries into digital media theory. The journal provides a publication environment for interdisciplinary research approaches, contemporary theory developments and methodological innovation in digital media studies. It invites reflection on how culture unfolds through the use of digital technology, and how it conversely influences the development of digital technology itself.

Board:
Maria Bakardjeva (University of Calgary), David Berry (University of Sussex), Jean Burgess (Queensland University of Technology, Brisbane, Australia), Mark Coté (King's College London), Colin Cremin (University of Auckland), Sean Cubitt (Goldsmiths, University of London), Mark Deuze (University of Amsterdam), José van Dijck (Utrecht University), Delia Dumitrica (Erasmus University Rotterdam), Astrid Ensslin (University of Alberta, Edmonton), Sonia Fizek (Abertay University), Federica Frabetti (University of Oxford), Orit Halpern (The New School, New York), Irina Kaldrack (Braunschweig University of Art / Leuphana University of Lüneburg), Denisa Kera (National University of Singapore), Lev Manovich (The Graduate Center, The City University of New York), Janet H. Murray (Georgia Institute of Technology, Atlanta), Jussi Parikka (University of Southampton), Lisa Parks (University of California, Santa Barbara), Dominic Pettman (The New School, New York), Rita Raley (University of California, Santa Barbara), Richard Rogers (University of Amsterdam), Julian Rohrhuber (Robert Schumann School of Music and Media, Düsseldorf), Marie-Laure Ryan (University of Colorado, Boulder), Mirko Tobias Schäfer (Utrecht University), Jens Schröter (University of Bonn), Trebor Scholz (The New School, New York), Tamar Sharon (Maastricht University), Roberto Simanowski (City University of Hongkong), Nathaniel Takcz (University of Warwick), Geoffrey Winthrop-Young (University of British Columbia, Vancouver), Sally Wyatt (Maastricht University)

Eduardo Luersen is a NOMIS Fellow at eikones, Center for the Theory and History of the Image at the University of Basel, and Associated Fellow at the Zukunftskolleg, University of Konstanz.
James Wilson is an Assistant Professor of Medieval History at the University of Groningen and Associated Fellow at the Zukunftskolleg, University of Konstanz.

Eduardo Luersen, James Wilson (eds.)

Digital Culture & Society (DCS)

Vol. 11, Issue 1/2025 – Digital Games through Muddled Pasts
and Modded Histories

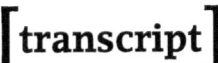

Bibliographic information published by the Deutsche Nationalbibliothek
The Deutsche Nationalbibliothek lists this publication in the Deutsche Nationalbibliografie; detailed bibliographic data are available in the Internet at https://dnb.dnb.de

Indexiert in EBSCOhost-Datenbanken.

© **2025 transcript Verlag, Bielefeld**
transcript Verlag | Hermannstraße 26 | D-33602 Bielefeld | live@transcript-verlag.de

All rights reserved. No part of this book may be reprinted or reproduced or utilized in any form or by any electronic, mechanical, or other means, now known or hereafter invented, including photocopying and recording, or in any information storage or retrieval system, without permission in writing from the publisher.

Cover concept: Kordula Röckenhaus
Printed by: Elanders Waiblingen GmbH, Germany
Print-ISBN 978-3-8376-6869-8
PDF-ISBN 978-3-8394-6869-2
ISSN of journal: 2364-2114
eISSN of journal: 2364-2122

Printed on permanent acid-free text paper.

Content

Digital Games through Muddled Pasts and Modded History
Introduction
Eduardo Luersen and James Wilson   7

I   Producing Historical Games and Heritage Experiences

Source Material and the Problem of Authenticity in Historical Game Development
William Hepburn and Jackson W. Armstrong   23

Historians Making Games
Unveiling a History Game Design Ethos
Magnus Henrik Sandberg, Eirik Brazier, and Ragnhild Hutchison   41

Enchanted Imaginings
Involving Museum Visitors in Heritage Adventures
Eva Kingsepp and Linda Ryan Bengtsson   63

II   Culture and Emotions as Historical Game Design Affordances

Cultural Combinatorics and Conjured Spectres
The Representation of Culture and Cultural Hybridity through the Game Mechanics of *Crusader Kings III*
Michael A. Conrad   85

Playing Adewale
The Politics of History in *Assassin's Creed's Freedom Cry*
Osvaldo Cleger   113

Historical Empathy and Player Agency in Computer Roleplaying Games
Kingdom Come: Deliverance and Pentiment
Robert Houghton   137

III   Revisioning, Reframing, Coding the Past

The Truth(iness) is a Lie
Historical Re-visions of the Cold War through
Call of Duty Paratexts
victoria l. braegger and Samantha Blackmon   163

Queering Hong Kong
Modded History in A Summer's End: Hong Kong 1986
Diego Barroso Sánchez   187

Calculated Actions
How Game Code Makes Arguments About the Past
James Baillie   211

Re-Enacting 9th Century Baghdad
Interview on the Narrative and Worldbuilding Aspects
of the Past, as Rendered in Assassin's Creed Mirage
James Wilson, Eduardo Luersen, Raphaël Weyland,
and Sarah Beaulieu   231

Biographical Notes   241

Acknowledgements   243

Digital Games through Muddled Pasts and Modded History
Introduction

Eduardo Luersen and James Wilson

"My men do not fear death, [...] they welcome it and the rewards it brings." This quote is how players of *Assassin's Creed* (Ubisoft 2007) are first introduced to the so-called 'leap of faith', the iconic gameplay mechanic and navigational element in which characters jump from implausible heights, before landing unharmed in carts filled with hay. This action has evolved into a signature feature of the franchise, which encourages players to climb culturally and architecturally significant buildings to reveal more information about the surrounding area. Yet many players may not realise that the episode represented in the opening scene of this hugely popular game is adopted almost verbatim from a 13th-century Old French chronicle (Daftary 1990: 6). Similarly, *Assassin's Creed: Mirage* (Ubisoft 2023), set in 9th-century Baghdad, actively engages with controversial historical subjects such as the 'Islamic Golden Age', the Zanj Rebellion, and the 'translation movement' from Greek to Syriac and Arabic that was patronised by the Abbasid Caliphate.

These and many other tropes in digital games raise important questions about how historical imaginaries are constructed for the medium. Such inquiries include the extent to which game developers seek to recover, select, update, and dramatise multifaceted and contested aspects of the past. So-called 'serious games' have been traditionally designed for education and training purposes across disciplines, and have embraced the pedagogical potential of games since at least the 1970s (Abt 1987 [1970]), with the term gaining traction in the 1980s and early 2000s (Susi/Johannesson/Backlund 2007; Djaouti et al. 2011). The assumption that games and play can serve educational purposes is traceable in intellectual traditions from Plato to Piaget (Wilkinson 2016), and it has supported the development of serious games while fuelling key tensions in the study of gamification, especially in the face of the increasing extrapolation of its promises – to the point of evangelism (Fuchs 2015). At the same time, discussions about the seriousness of games, game-based learning, and the workification of games lead to considerations about the nature of playing itself, as well as its specific situations and broader epistemic shifts (Ochsner et al. 2023). In this context, the question arises: what are the implications of drawing upon historical knowledge in game-like formats and playful environments?

Bringing these questions into the context of today's digitisation of culture and knowledge practices, this special issue of *Digital Culture & Society* addresses how knowledge about the past is crafted and curated in and for digital games. Beyond developing deliberate visions of the past through narrative design and game imagery, digital games make historical material accessible to wider audiences while becoming themselves materials for scholarship and public history. As multisensory media driving public engagement, contemporary history-themed games have been likened to a form of historical tourism (Schwarz 2024). Accordingly, digital games are often expected to serve as carriers for historical interpretations, incorporating contemporary political debates (Pfister/Görgen 2020), power structures (Seiwald/Wade 2023), and forms of cultural diplomacy (Donald/Webber/Wright 2023) into the historical settings they represent. This is not a challenge unique to games. Other mass media – from literature to cinema – have grappled for a long time with similar issues when depicting historical events and figures. However, digital games require us to reevaluate and update our understanding of the modalities of historical representation to include computer simulation, algorithmic rule structures, and user-oriented media mechanics, which interface knowledge about the past with operational structures such as narrative architecture (Jenkins 2004), procedural rhetoric (Bogost 2007), affective play (Jagoda/McDonald 2018). This means that while games may eventually exhibit similarities with traditional institutional structures, pedagogical methods, and monumental approaches, they also diverge in ways that shape how history and heritage, both tangible and intangible, are accessed and understood (Mochocki 2021; Houghton 2023), structuring knowledge mainly around numerous media affordances and genre conventions.

Given these structural affinities and differences, it is also unsurprising that the simulations of the past in games and their branding (Wright 2023) are shaped in relation to conventional – although highly contingent – aesthetic categories such as authenticity (Zimmermann 2020) or realism (Krapp 2019). This issue naturally extends beyond games that refer to history. Semiotic analyses of film and photography have long emphasised the indexical relationship of image and referents (Wollen 2013 [1969]), and while digital media alters this connection, indexicality remains influential in digital aesthetics and the meaningful experience of game worlds (Švelch 2008, Fernández-Vara 2011). Traditional audiovisual techniques such as motion capture, photogrammetry, and foley, contribute to modelling game worlds that look plausible and feel relatable. In games that rely on history or cultural memory specifically, developers may use archival documents, city planning blueprints, art historical motifs, or oral testimonies to build environments loosely or strictly inspired by historical work, but which ultimately may also resonate with extemporaneous aesthetic conventions and, ultimately, the non-coherent messiness that defines games (Bogost 2009). Realism in particular has attracted sustained interest from the games industry and game studies scholarship. As Peter Krapp (2019) points out, games across numerous genres have laid claim to different forms of realism: from offering situations that refer to recent news headlines to depicting the skills of celebrity athletes, from simulating the driving experience of classical cars, the physical ballistics of weapons, and aircraft manoeuvres, to enacting details of military tactics and the complexities of global economic systems. Whether they deliberately refer to the past or not, the emphasis on verisimilitude remains a recurring aspect in game design and critique.

Concepts like authenticity also contain pitfalls when looked at from a historical perspective. The past decade has seen an impressive body of literature (Kapell/Elliott 2013; Wright 2022) that engages with the notion of 'historical accuracy' in digital games and their historiographical potential (Chapman 2016). As editors of this special issue, we have chosen to sidestep this debate somewhat, although we of course acknowledge that historical accuracy is a convoluted concept, which is dependent upon the nature of the sources one relies upon to interact with any given historical figure, setting, event or process. What we judge to be more pertinent are the ways in which those involved in game development navigate the complexity of these ideas, while simultaneously attempting to promulgate the historical *bona fides* of their developmental process. The standard practice for achieving this is through visual representations of art, architecture or material objects, the use of narrative structures, the creation of credible characters and the involvement of external experts (Schwarz 2020). Additionally, a desire to apply a narrow sense of historical or cultural 'accuracy' to specific historical settings can paradoxically lead to a perception of inauthenticity from players, who may hold their own preconceptions about the past that no longer conform to the current state of the art (Burgess/Jones 2022; Slingluff et al. 2024: 481).

Beyond debates about the credibility of historical representations, another frequent question addresses how simulations of the past are experienced by players (Kapell/Elliott 2013; Caselli/Bonello Rutter Giappone/Majkowski 2023), foregrounding their affective responses, which suggests that games formally enact a sense of mediation with the past. Moreover, as digital games deliberately draw from literature, theatre or film, but also from historical research (and, sometimes, though less commonly, from historical methods), understanding how this technically mediated affection is engendered becomes increasingly important. In a formal sense, the seeming experience of playing with the past is achieved by building additional layers of mediation, as with other media. From a *remediation* outlook, Jay David Bolter and Richard Grusin (1999) described this process in the late 1990s as a dynamic of hypermediation: counter-intuitively, the artifice of immediacy is produced through the deliberate multiplication of different media layers. More broadly, but still in a strictly structural formal setting, film-maker and theorist Sergei Eisenstein unconventionally likened the development of early film form, in particular the process of *montage*, to airplane design: engineers – said Eisenstein to his students at the Gerasimov Institute of Cinematography in Moscow (Avellar 2002) – failed to make machines fly when mimicking the exact form of bird wings, but succeeded when decomposing their multiple functions (propulsion, lift, gliding), recombining them into new configurations. While the analogy, drawing from an old-fashioned functionalist principle of modernism, may seem largely anecdotal, the way by which digital games imbue a feeling of presence to the imagined past is, to a certain extent, relatable to this process. Games may struggle to find their place when replicating the aesthetic paradigms of preceding media, but they do draw on their latent repertoire to advance the medium's own formal logics. Let alone the deliberate fusion of narrative and playful forms in the intermedial textuality of 21th-century audiovisual media (Schubert 2019), one can consider that formal experimentations with the medium of digital games feature a standing reserve for extending the possibilities of presenting the findings of historical research and, perhaps more interestingly, other forms to reflect on its methods.

To avoid letting this interest in the particular – or, more precisely, the *potentially distinct* – affordances of historical games lead us into the familiar rhetoric of perpetual novelty that often surrounds the medium (Luersen 2020; Unterhuber 2024), which makes the very history of digital games itself be read as "hagiographic" (Lawler/Smith 2021: 50) and "unable to relate […] to wider cultural framework(s)" (Huhtamo 2005: 5), we should also acknowledge the need for a broader perspective. Numerous contributions, including entire edited collections, have been dedicated to exploring these and many other dimensions of the relationship between games, historical knowledge, and learning experiences (cf. Whalen/Taylor 2008; Chapman 2016; Mai/Preisinger 2020; Ariese-Vandemeulebroucke et al. 2021; Preisinger 2021; McCall 2022; Houghton 2022; Schwarz 2023; Champion/Vera 2023). Considering the growing body of literature in the niche of historical game studies, important aspects that remain substantially under-explored are the actual settings in which the simulations of the past are produced, with the social studies of game production (Sotamaa/Švelch 2021) and game-making contexts (Grufstedt 2022) having so far received comparatively little attention. By observing how narrative designers, developers, and historians collaborate, we can gain insights into the handling of knowledge about the past in game design, especially when the pursuit for a sense of authenticity is conflated with the demands of user-oriented digital media. No matter how attached to a given source developers may hope to be, "staffing considerations, time, resource management, efficiency, and the technological complexities of the medium will affect what can be accomplished" (Slingluff et al. 2024: 481). Cultural representations are thus embedded in a web of formal, technical, ideational, and socio-economic factors, with the outcomes resulting from decisions but also entangled practical – when not accidental – aspects that affect how the game worlds are modelled from the onset and throughout the development process.

Despite the potentially significant impact of these circumstances, the mundane production routines and technical design decisions that affect how the past is represented in digital games have often been secondary to debates about these artefacts. With this special issue we wish to draw attention to how these games are produced and how they articulate knowledge about the past with the labour of designing games, while also contributing to the established body of research on the representations of the past in digital games, their educational potential and limitations, as well as the broader imaginaries of history that circulate in the games industry. How do game production aspects and the aesthetic conventions of games affect the relationship established with the past, with historical records, or with previously established historical imaginaries? On the other hand, how can the archival materials shape the creative processes of game design? To pursue these questions, we cannot just interpret the representations of the past; we need to understand the sites of their creation. By scrutinising design features and affordances, the cultural values at stake in game production, practical workflows, and developer decisions surrounding political and aesthetic elements, one can gain constructive insight into often-overlooked dynamics that influence how visions of the past are crafted in and for digital games. While not all the contributions of this issue address this topic directly, and the body of ongoing work within this niche still remains very limited, we expect that this issue can contribute to advancing the understanding of these dynamics, stimulating further research endeavours in this direction.

Beyond the sites of game-making, and in light of the contemporary convergence between game production, consumption and the platform economy (Young 2021; Thorhauge 2023), the games and history nexus should also be aware of how historical knowledge may be employed as raw material not only in development but also within the prevalent hypercirculation of images, or their general 'iconomy' – as suggested with Peter Szendy's (2019) playful portmanteau. This is an important framework to consider how historical tropes may also happen to be commodified by the games industry, through the dedicated monetisation expertise (Roessel/Švelch 2021; Koenig et al. 2024) integrated within the wider ecology of platforms. The present-day accelerated economy of images suggests that images circulate not only as aesthetic or representational objects, but as units of valuable data within the global visual economy. Within the regimes of visibility shaped by global platforms and media infrastructures, images (game assets included) are not just developed, consumed, and interpreted, but counted, exchanged, and organised as commodified tokens and items for expanded micro-transactions (Thorhauge 2024) that can be algorithmically stored, managed, and monetised. In the context of this issue, it matters how images of the past may circulate as part of a composite patchwork of media mash-ups (Cole 2022), quite loosely remixed throughout digital culture. In this milieu, historical representations are conveyed as assets that operate within a broader digital attention economy, with gaming platforms, among other service-based applications, drawing from circulating visual materials. Throughout the modular reuse of images shaped as 'historical content', data-driven media may recycle and recombine visuals from different times and places, frequently detached from their contexts, shaping cultural tropes according to the logics of platform algorithms (Es/Verhoeff 2023), with or without human involvement in the process (Paglen 2016). In this context, the circulation value and the rate of exchange of images, overshadows images themselves. As a result, game images of the past may be valued less for their visuality or representational estimation *per se* than for the value they contribute to the loop of accelerated circulation.

In light of these developments, this special issue of *Digital Culture & Society* considers subjects that arise across multiple dimensions of the intermingling between digital games and history. We do so by being open to different branches of research, spanning from representational aspects, paratextual analysis, methodological contributions, practices of game production, and political-aesthetical implications of historical games. This endeavour is led by a broad set of questions at the interplay of game design and historical knowledge, encompassing the heterogeneous ways authors in this special issue contribute to its general scope: How are the imaginaries of in-game historical conflicts or cross-cultural collaboration developed? How does the classification of digital games into different genres influence how we analyse historical aspects? In which ways are history-based games serious? How are game logics and structures used as tools for historical engagement by organisations? What are the different implications of digital games for the reception of historical knowledge when history is meant to be played as a user-oriented medium? What can be gained from analyses of paratexts, and which insights can they provide into these processes? How is the decision-making dynamic in historical games tailored between developers, concept designers, and historical advisors? How do game studio workflows play a role in shaping how historical games are developed? What interdisciplinary methods can be developed to study the intersection of digital games and historical knowledge?

Engaging with this wide spectrum of questions, we divide this issue into three main axes: (1) *Producing Historical Games and Heritage Experiences,* (2) *Culture and Emotions as Historical Game Design Affordances,* (3) *Revisioning, Reframing, Coding the Past.*

The first section opens with the paper *Source Material and the Problem of Authenticity in Historical Game Development,* by **William Hepburn** and **Jackson W. Armstrong**. The paper addresses a key question: how one of the scholarly foundations of historical research – namely, the use of primary sources – can meaningfully inform game design and be integrated into game-making practices. Drawing on the authors' experience developing the game *Strange Sickness* (Common Profyt Games 2021), the article examines challenges around historical game development, in particular the use of historical sources, and argues that embracing the interpretive nature of historical scholarship can reconcile the differing aims of historians and developers.

The second paper in this section is *Historians Making Games: The Unveiling of a History Game Design Ethos,* in which **Magnus Henrik Sandberg, Eirik Brazier,** and **Ragnhild Hutchinson** describe and analyse design production values of *The Widow's Boutique* (Tidvis 2023). The visual novel game integrates primary source materials and historical research into its narrative structure, relying on a rich body of historical documentation. While focusing on how the game studio's interest and the developers' scholarly research on cultural heritage informs design values, the paper unfurls practical dilemmas and raises questions about the particular stakes distinguishing the development of games portraying actual historical settings from openly fictional game design.

In their article *Enchanted Imaginings: Involving Museum Visitors in Heritage Adventures*, **Eva Kingsepp and Linda Ryan Bengtsson** explain the site-specific curation of playful physical objects and augmented reality applications they developed for a heritage site in the mining area of Långban, Sweden. While detailing modelling principles and design decisions that they made concerning their source materials, their contribution also emphasises the importance of embracing the role of participants as co-producers of physically situated and embodied interpretations of the past.

Opening the second section, **Michael A. Conrad** analyses the cultural system and the introduction of cultural hybridity mechanics in *Crusader Kings III* (Paradox Development Studio 2020). In *Cultural Combinatorics and Conjured Spectres: The Representation of Culture and Cultural Hybridity through the Game Mechanics of Crusader Kings III*, he proposes a rich methodological approach, combining the analysis of the in-game representation of cultural data with the revision of developer diaries documenting design discussions and collaborative websites and forums utilised by the community of players. With this, the paper unfurls the mechanics and other abstract dynamics structuring how the game's rendition of the past, including its system of cultural hybridity, is set up to be experienced by players.

The second contribution, *Playing Adewale: The Politics of History in Assassin's Creeds Freedom Cry*, describes how empathy-driven mechanics and the narrative framing of *Assassin's Creed Freedom Cry* (Ubisoft 2013) constructs an aesthetic continuum between the context of the Haitian Revolution and contemporary Black civil rights struggles. By analysing design aspects, narrative and audiovisual tropes, as well as player reception, **Osvaldo Cleger** draws attention to how this game diverges with the signature monumental and touristic historical representations of the Assassin's Creed franchise, foregrounding a politically charged dynamic of slavery and resistance in 18th-century Saint-Domingue.

The third paper in this section is *Historical Empathy and Player Agency in Computer Roleplaying Games: Kingdom Come: Deliverance and Pentiment*, by **Robert Houghton**. The article explores the strategies games employ when engaging players through emotional identification and role-playing, rather than through realism or intricate systemic simulation. It argues that the core conventions of roleplaying games, particularly the player agency they afford, can create open opportunities for developing a nuanced understanding of historical situations and experiences, inviting players to relate to the perspectives of historical characters.

Opening the last section is the paper *The Truth(iness) is a Lie: Historical Re-Visions of the Cold War through Call of Duty paratexts*. By exploring the promotional paratexts of *Call of Duty: Black Ops 6* (Treyarch/Raven Software 2024),

victoria l. braegger and **Samantha Blackmon** employ the satirical notion of 'truthiness' to describe the emotional appeal mobilised in the marketing strategy prior to the release of the game. braegger and Blackmon argue that, by merging real and fictionalised media, mixing archival footage, fabricated references, and anecdotal cultural references, *Call of Duty* seeks to construct its own fabricated version of Cold War history, resonating strongly with contemporary political aesthetics.

In the following contribution, *Queering Hong Kong: Modded History in A Summer's End: Hong Kong 1986*, **Diego Barroso Sánchez** also analyses a visual novel game. Articulating hermeneutical cultural analysis and interviews with developers, the paper describes how *A Summer's End: Hong Kong 1986* (Oracle and Bone 2020) reframes the cultural-historical memory surrounding the handover of Hong Kong from British colonial rule to the People's Republic of China in 1997, particularly in relation to colonial legacies and queer perspectives. Contributing to a growing interest in how titles delivered by smaller studios participate in historical representation, the analysis affords the game with a speculative engagement with the imagined past, in an openly optimistic setting, at the backdrop of personal and political uncertainty.

In *Calculated Actions: How Game Code Makes Arguments About the Past*, **James Baillie** explores how digital games structure the ways in which players interact with historical or pseudo-historical worlds through underlying code and mechanics. Moving beyond the surface layer, the paper eloquently unfurls how game models mediate the actionable possibilities within the simulations, thereby embedding historiographical conceptions into their procedural logic. In doing so, the article contributes the epistemology of game design by showing how the mediation of games enacts implicit understandings of the past through the assumptions, aesthetics and actions modelled in several game worlds.

After the thematic sections, and completing this special issue, we include an interview with **Raphaël Weyland**, historian at Ubisoft Montréal, and **Sarah Beaulieu**, narrative designer for *Assassin's Creed Mirage* (Ubisoft 2023). The interview explores some of their visions on the role of historical subjects in the narrative structure and world building of digital games, as well as their own insight into production aspects of the gaming industry projects they have been involved with. The perspectives shared in this interview are those of the developers and do not reflect the views of the editors. We hope this conversation contributes to analyses of the production dynamics and stakes involved in games that engage with the past and historical sources. This contribution is a transcription of the interview they conceded to us for the *Digital Games through muddled pasts and modded history* workshop, organised in April 2024 at the Zukunftskolleg at University of Konstanz. We thank them both for their interest and willingness to participate in the workshop and take part in the interview, and to Ubisoft Divertissements Inc. for granting us the permission to publish it.

We want to thank the authors of the papers included in this special issue for their valuable contributions. We would also like to express our thanks to all the anonymous peer reviewers for their time, generous feedback and constructive evaluations. We are also grateful to the chief editors of Digital Culture & Society, and in particular to Mathias Fuchs, whose support was central in bringing this issue to fruition. We would also like to acknowledge the support of the Zukunftskolleg at the University of Konstanz, where the two of us shared an office for the past three years. The institute provided not only structural and financial support, but also an intellectually stimulating environment that enabled us to organise the aforementioned workshop on this topic – an endeavour that eventually led to the development of this special issue. We are likewise grateful to the GameLab at the university and Benjamin Schäfer for their support in the context of the workshop and related events. We also thank Beate Ochsner and the Serious Gaming initiative at the University of Konstanz for invaluable discussions on games in their epistemological, socio-historical, imaginary, and ecological dimensions, including subjects closely related to the scope of this publication. We believe this special issue genuinely represents the result of a commitment to fostering interdisciplinary collaboration, including the constructive acceptance of the attendant risks. This cooperation was funded by the Federal Ministry of Education and Research (BMBF) and the Baden-Württemberg Ministry of Science as part of the Excellence Strategy of the German Federal and State Governments.

References

Abt, C. C. 1987 [1970]: Serious Games. Boston: University Press of America.

Ariese-Vandemeulebroucke, C. E./Boom, K. H. J./dem Hout, B. v./Mol, A. A. A./Politopoulos, A. (2021): Return to the Interactive Past: The Interplay of Video Games and Histories. Leiden: Sidestone Press.

Avellar, J. C. (2002): "Introdução: E = mc². " In: S. Eisenstein, A Forma do Filme. Rio de Janeiro: Jorge Zahar.

Bogost, I. (2007): Persuasive Games: The Expressive Power of Videogames. Cambridge: MIT Press.

Bogost, I. (2009). "Videogames are a Mess. My DiGRA 2009 Keynote on Videogames and Ontology" (3 September 2009). Retrieved from http://www.bogost.com/writing/videogames_are_a_mess.shtml.

Bolter, J. D./Grusin, R. (1999): Remediation: Understanding New Media. Cambridge: MIT Press.

Burgess, J./Jones, C. (2022): Exploring Player Understandings of Historical Accuracy and Historical Authenticity in Video Games. Games and Culture 17(5), pp. 816–835.

Caselli, S./Bonello Rutter Giappone, K./Majkowski, T. Z. (2023): Ten Years of Historical Game Studies: Towards the Intersection with Memory Studies. Game 10, pp. 29–50.

Champion, E./ Vera, J. H. (2023): 'Assassin's Creed' in the Classroom: History's Playground or a Stab in the Dark. Berlin: De Gruyter.

Chapman, A. (2016): Digital Games as History: How Videogames Represent the Past and Offer Access to Historical Practice. London: Routledge.

Cole, R. A. (2022): Mashing Up History and Heritage in Assassin's Creed Odyssey. Games and Culture 17(6), pp. 915–928.

Common Profyt Games Ltd (2021): Strange Sickness [Video Game]. Common Profyt Games Ltd

Daftary, F. (1990): The Isma'ilis: Their History and Doctrines. Cambridge: Cambridge University Press.

Djaouti, D./Alvarez, J./Jessel, J./Rampnoux, O. (2011): "Origins of Serious Games." In: M. Ma/A. Oikonomou, L. C. Jain (eds.), Serious Games Edutainment Applications. Berlin: Springer, pp. 25–43.

Donald, I./Webber, N./Wright, E. (2023): Video Games, Historical Representation and Soft Power. Journal of Gaming & Virtual Worlds 15(2), pp. 105-127.

Es, K. v./Verhoeff, N. (2023): Situating Data: Inquiries in Algorithmic Culture. Amsterdam: Amsterdam University Press.

Fernández-Vara, C. (2011): "Game Spaces Speak Volumes Indexical Storytelling." In: Proceedings of DiGRA 2011 International Conference: Think Design Play. Hilversum, September 14–17, Conference Proceedings.

Fuchs, M. (2015): "Total Gamification: Introduction." In: M. Fuchs (ed.), Diversity of Play. Lüneburg: Meson Press, pp. 7–20.

Grufstedt, Y. (2022): Shaping the Past: Counterfactual History and Game Design Practice in Digital Strategy Games. Berlin: De Gruyter.

Houghton, R. (2022): Teaching the Middle Ages through Modern Games: Using, Modding and Creating Games for Education and Impact. Berlin: De Gruyter Oldenburg.

Houghton, R. (2023): Awesome, but Impractical? Deeper Engagement with the Middle Ages through Commercial Digital Games. Open Library of Humanities 9(2), n.p.

Huhtamo, E. (2005): "Slots of Fun, Slots of Trouble: An Archaeology of Arcade Gaming." In: J. Raessens/J. H. Goldstein (eds.), Handbook of Computer Game Studies. Cambridge: MITPress, pp. 3–21.

Jagoda, P./McDonald, P. (2018): "Game Mechanics, Experience Design, and Affective Play." In: J. Sayers (ed.), Routledge Companion to Media Studies and the Digital Humanities. New York: Routledge, pp. 174–182.

Jenkins, H. (2004): "Game Design as Narrative Architecture." In: N. Wardrip-Fruin/P. Harrington (eds.), First Person: New Media as Story, Performance, and Game. Cambridge: MIT Press, pp. 118–130.

Kapell, M./Elliott, A. B. R. (2013): Playing with the Past: Digital Games and the Simulation of History. London: Bloomsbury Academic.

Koenig, N./Denk, N./Pfeiffer, A./Wernbacher, T./Wimmer, S. (2024): Money | Games | Economies. Krems: University of Krems Press.

Krapp, P. (2019): "Sid Meier's Civilization. Realism." M. T. Fayne/N. B. Huntemann (eds.), How to Play Video Games. New York: New York University Press, pp. 44–51.

Lawler, J./Smith, S. (2021): Reprogramming the History of Video Games: A Historian's Approach to Video Games and Their History. International Public History 4(1), pp. 47–54.

Luersen, E. (2020): "Ressonância Tecnocultural: Rastros da Ambiência Contemporânea nas Sonoridades dos Jogos Digitais" [Doctoral thesis, Unisinos University], http://www.repositorio.jesuita.org.br/handle/UNISINOS/9316.

Mai, S. F./Preisinger, A. (2020): Digitale Spiele und historisches Lernen. Frankfurt: Wochenschau.

McCall, J. B. (2022): Gaming the Past: Using Video Games to Teach Secondary History. London: Routledge.

Mochocki, M. (2021): Heritage Sites and Video Games: Questions of Authenticity and Immersion. Games and Culture 16(8), pp. 951–977.

Mochocki, M. (2022): Editorial: Games with History, Heritage, and Provocation. Games and Culture 17(6), pp. 839–842.

Ochsner, B./Willkomm, J./Waldrich, H./Spöhrer, M. (2023): 'Serious Gaming' – oder Spielen ernst nehmen: Ein Forschungsprogramm. Zeitschrift für Medienwissenschaft 15(1), pp. 123–136.

Oracle and Bone (2020): A Summer's End: Hong Kong 1986 [Video Game]. Oracle and Bone.

Paglen, T. (2016): "Invisible Images (Your Pictures Are Looking at You)" (8 December 2016), Retrieved from https://thenewinquiry.com/invisible-images-your-pictures-are-looking-at-you/.

Paradox Development Studio (2020): Crusader Kings III [Video Game]. Paradox Interactive.

Pfister, E./Görgen, A. (2020): "Politische Transfer prozesse in digitalen Spielen: Eine Begriffsgeschichte." In: A. Görgen/S. H. Simond (eds.), Krankheit in digitalen Spielen. Bielefeld: transcript, pp. 51–74.

Preisinger, A. (2021): Digitale Spiele in der historisch-politischen Bildung. Frankfurt: Wochenschau

Roessel, L. v./Švelch, J. (2021): "Who Creates Microtransactions: The Production Context of Video Game Monetization." In: O. Sotamaa/J. Svelch (eds.), Game Production Studies. Amsterdam: Amsterdam University Press, pp. 197–215.

Schubert, S. (2019): "Narrative and Play in American Studies: Ludic Textuality in the Video Game Alan Wake and the TV Series Westworld." In: S. Pöhlmann. Playing the Field: Video Games and American Studies. Berlin: De Gruyter, pp. 113–130.

Schwarz, A. (2022): "History in Video Games and the Craze for the Authentic." In: M. Lorber/F. Zimmermann (eds.), History in Games: Contingencies of an Authentic Past. Bielefeld, transcript, pp. 117–137.

Schwarz, A. (2023): Geschichte in digitalen Spielen: Populäre Bilder und historisches Lernen. Stuttgart: Kohlhammer.

Schwarz, A. (2024): "Discovering the Past as a Virtual Foreign Country: Assassin's Creed as Historical Tourism." In: E. Champion/J. F. H. Vera (eds.), Assassin's Creed in the Classroom: History's Playground or a Stab in the Dark? Berlin: De Gruyter, pp. 169–187.

Seiwald, R./Wade, A. (2023): A Genealogy of Power: The Portrayal of the US in Cold-War Themes Videogames. Journal of the Austrian Association for American Studies 4(2), pp. 270–291.

Slingluff, S./Vural, D./Anderson, G. D./Etefaghi. D. (2024): The 200-Million Student Classroom: Teaching Islamicate History, One Video Game at a Time? International Journal of Middle East Studies 56(3): pp. 465–484.

Sotamaa, O./Švelch, J. (2021): Game Production Studies. Amsterdam: Amsterdam University Press.

Susi, T./Johannesson, M./Backlund, P. (2007): "Serious Games – An Overview." Technical Report HS- IKI -TR-07-001. University of Skvde: Skvde School of Humanities and Informatics.

Švelch J. (2008): "What you Can't See is What You Don't Get: Paradigms of Game World Visualization." In: Proceedings of the 2008 Conference on Future Play: Research, Play, Share (03–05 November 2008), Toronto, pp. 212–215.

Szendy, P. (2019): Supermarket of the Visible: Toward a General Economy of Images. New York: Fordham University Press.

Thorhauge, A. M. (2023): Games in the Platform Economy: Steam's Tangled Markets. Bristol: Bristol University Press.

Thorhauge, A. M. (2024): The Steam Platform Economy: From Retail to Player-Driven Economies. New Media and Society 26(4), pp. 1963–1983.

Tidvis (2023): The Widow's Boutique [Video Game]. Tidvis.

Treyarch/Raven Software (2024): Call of Duty: Black Ops 6 [Video Game]. Activision Blizzard.

Ubisoft (2007): Assassin's Creed [Video Game]. Ubisoft.

Ubisoft (2013): Assassin's Creed Freedom Cry [Video Game]. Ubisoft.

Ubisoft (2023): Assassin's Creed: Mirage [Video Game]. Ubisoft.

Unterhuber, T. (2024): "Das ewige neue Medium: Die Geschichtslosigkeit der Computerspielgeschichte." In: T. Spies/F. Seyda/H. Pötzsch (eds.), Spiel*Kritik. Bielefeld: transcript, pp. 107–122.

Whalen, Z./Taylor, L. N. (2008): "Playing the Past: An Introduction." In: Z. Whalen/L. N. Taylor (eds.), History and Nostalgia in Video Games. Nashville: Vanderbilt University Press.

Wilkinson, P. (2016): Brief History of Serious Games: Lecture Notes in Computer Science. Entertainment Computing and Serious Games 9970, pp. 17–41.

Wollen, P. (2013 [1969]): Signs and Meaning in the Cinema. London: British Film Institute.

Wright, E. (2023): Still Playing with the Past: History, Historians and Digital Games. History & Theory 61(4), pp. 166–177.

Wright, E. (2023): "Paratexts, 'Authenticity', and the Margins of Digital (Game) History." In: R. Seiwald/E. Vollans (eds.), (Not) In the Game: History, Paratexts, and Games. Berlin: De Gruyter Oldenbourg, pp. 33–54.

Young, C. J. (2021): "Unity Production: Capturing the Everyday Game Maker Market." In. O. Sotamaa/J. Švelch (eds.), Game Production Studies. Amsterdam: Amsterdam University Press, pp. 141–158.

Zimmermann, F. (2020): "Approaching the Authenticities of Late Modernity." In: M. Lorber/F. Zimmermann (eds.), History in Games: Contingencies of an Authentic Past. Bielefeld: transcript, pp. 9–23.

Producing Historical Games and Heritage Experiences

Source Material and the Problem of Authenticity in Historical Game Development

William Hepburn and Jackson W. Armstrong

Abstract

History is a prevalent subject in digital games and a thriving interdisciplinary field of historical games studies has emerged in response. While there is communication between historians and game developers in the creation of historical games and in the study of historical games, Robert Houghton (2024) has noted that the "skillsets of game developers and historians are in many respects wildly different, and academics are typically not well placed to lead the development of a game", and that the goals of the two groups in their approach to history are widely divergent: academic historians focussed on "expressing a scholarly historical account" and game developers focussed on how history can be used to create a more compelling game, usually driven by the pressure of commercial viability (47). The growth of independent game development culture and the increasing accessibility of game development tools with relatively low technical barriers to entry in the last two decades has provided opportunities for academic historians to close this gap (Juul 2019; Anthropy 2012). Indeed, there is a growing literature reflecting on attempts by academic historians to do so (see Reid 2024), to which this article is a contribution. The discussion which follows here is based in part on the authors' experience as academic historians leading the development of a historical game called Strange Sickness. The article begins by outlining the problem of authenticity in historical games in relation to the development of historical theory. It follows with an account of the game's development and its relationship to historical sources and academic research on those sources. Finally, the ways in which primary sources are used and referenced in Strange Sickness and other games are discussed. We argue that the apparent gulf in objectives between game developers and historians can be bridged to their mutual benefit by embracing the authored, contingent nature of historical interpretation of primary sources, which reflects historical practice while offering a form of authenticity with appeal to game playing audiences.

Keywords

videogames, history, interpretation, authenticity, medieval, Aberdeen, Scotland.

Historical theory and the problem of authenticity in historical games

The study of how games can be historical is the domain of historical game studies. In the last two decades this field has emerged with its own identity, approaches and thriving scholarly community (Chapman et al 2016: 358–359). It is defined by Adam Chapman (2016), one of its chief practitioners, as "the study of those games that in some way represent the past or relate to discourses about it." (16) Here we are focussed on historical games that are, to some extent, "explicitly identified as real rather than imaginary" (White 1987: 3; Zucconi et al 2013: 199). This can include games set in fictionalised, or remixed, historical settings such as *Pentiment* (Obsidian Entertainment 2022), set around a fictional abbey in sixteenth-century Bavaria, drawing on historical referents but also a literary inspiration: Umberto Eco's *The Name of the Rose* (Kleinerman et al 2023: 2). Also in this category are the *Civilization* (2K Games 1991–2025) games, with their remixable kaleidoscope of events, people and places from the past, and games like *Kingdom Come: Deliverance* (Warhorse Studios 2018) which, while they include fictionalised elements, make real events, people and places from a specific place and time central to the game's narrative and setting (Cook 2024: 27–28).

This explicit identification as real can occur in various ways, such as through marketing. In the process games may make claims to historical authenticity but in so doing they often perpetuate a long-outdated epistemology of history. As history came to be established among other professional academic disciplines in the nineteenth century, its practitioners placed great emphasis on technical skills and sought to reveal the positivist empirical reality of the past. The twentieth century, by contrast, led to the intellectual trend of structuralism (which in history is best embodied in the *Annales* school) and in the later decades of the century the postmodernist linguistic turn. Hayden White, a key theorist of historical postmodernism, observed that history was made from language, or the "meaning-fraught formulations that emerge from the active mind of the historian-writer, and are not passively copied from the world" (Partner 2012: 7).

The underlying epistemologies of these broad scholarly phases are addressed in historical games studies, most notably in Chapman's (2016) application of Alan Munslow's framework of reconstructionist, constructionist and deconstructionist epistemologies – broadly connected, respectively, to the empiricist, structuralist and postmodernist approaches outlined above – to games (see also Clyde et al 2012: 7–8). For Chapman the key distinction in historical games is between realist simulations and conceptual simulations. A key part of this distinction is that in realist simulations history is primarily represented aesthetically, such as through a game's visuals, while in conceptual simulations the game's mechanics are the key means of representing the history. In Chapman's schema these represent, respectively, reconstructionist and constructionist epistemologies, the former focussed on historical truth independent of the subjective input of the historian, and the latter on the use of theories and concepts to determine the structures governing the past. Both are ultimately positivistic in their belief, through differing methods, in the recoverability of past reality (Chapman 2016: 59–89).

Despite some qualifications (see Houghton 2024: 59; Cruz Martínez 2020), Chapman's framing is still apt to describe prevailing epistemologies of history in historical games in the sense that, even when presenting structuralist historical arguments, games often assume that past reality can be reconstructed. This is reflected in scholarship on the concept of authenticity in historical games, which has emerged as an important theme in the work of scholars such as James Cook (2024) and, especially, Esther Wright (2018: 602; 2022). This scholarship addresses authenticity as "an agreed upon construct" about the past, and points to the need for further work in this area (Wright 2022: 177). Despite exceptions such as *Pentiment*, which is sensitive to the contingent nature of historical knowledge, claims to authenticity in games overwhelmingly reflect positivist epistemologies of history (Wright 2024: 191). These claims have a simple and popular appeal, in which the authenticity of a game is measured by its ability to capture past reality, its "historical accuracy". Indeed, as Jerremie Clyde et al. (2012) note, historians too have been complicit in framing the representation of the past in games through this lens when using games in teaching (4). This frames authentic history as a value-free effort to reveal the facts of the past when, in truth, what is presented as objective past reality is itself shaped by prevailing cultural ideas about a given period or subject. Karl Alvestad and Robert Houghton have noted this in the context of games about the Middle Ages (Houghton 2024: 22; Houghton/Alvestad 2022: 3). Deeply entrenched cultural ideas about the Middle Ages as a distinctly non-modern world, defined as a rejected past, and heavily shaped by nineteenth-century mythmaking, mean that medievalism, including in games, is an understanding of the period that does not depend on scholarship.

Claims of authenticity are a key part of the appeal of historical games explicitly identified as real. Authenticity is often "an element of branding" (Gundermann et al. 2025: 10) for historical games as it is in the *Assassin's Creed* (Ubisoft 2007–2025) series, which advertises opportunities to enter into history, echoing the animus from the in-game fiction which allows characters to travel from the modern day into the historical settings embodied in their ancestors' memories (Wright 2018: 602). Thus they create a time travel fantasy in which players overcome Marc Bloch's observation that "we have no other device for returning through time except that which operates in our minds with the materials provided by past generations" (Bloch 1992: 27). The concept of historical authenticity can be framed as the relationship between object authenticity, the adjudged authenticity of a given historical representation, and subject authenticity, the feeling or perception of authenticity (Gundermann et al 2025: 10). The problem of authenticity for an academic historian entering game development is that they find themselves facing the demand, as the marketing for *Assassin's Creed* illustrates, for a feeling of subject authenticity from players of historical games who are accustomed to a definition of object authenticity – conformance to a recoverable past reality – that, in the context of the present-day historical profession is outdated and naïve (Gundermann et al 2025: 11).

Strange Sickness: a historian-led indie game development project and its objectives

The perspective we bring to these issues comes from our development of *Strange Sickness* (Common Profyt Games 2021).[1] The game was grounded in our work as academic historians engaged in teaching and research in a university context, especially through the Aberdeen Burgh Records Project. This work, digital and interdisciplinary in nature but rooted in the study of historical source materials, began in 2012 as a cooperation between the Aberdeen City & Aberdeenshire Archives and the University of Aberdeen, and has encompassed several grant-funded collaborations[2] (see Aberdeen Registers 2016-). The focus of these sub-projects has been the investigation of municipal council register volumes from medieval Aberdeen, the handwritten records of local government which illuminate in detail aspects of the administration of this Scottish town, or burgh. Surviving in near-continuous sequence from 1398 onwards, these council registers were known at the time as "common books" (Hepburn/Small 2020). They functioned as a repository for a wide range of written information including the enactment of local ordinances, court proceedings, lists of taxpayers, the election of town officials, external correspondence, and so forth. The earliest eight of the surviving register volumes, covering the years 1398 to 1511, have been inscribed on the UNESCO UK Memory of the World Register (see UNESCO 2013), in recognition of their historical significance to Scotland and the United Kingdom. In 2019 *Aberdeen Registers Online: 1398-1511* was published as a digital edition (see ARO 2019), alongside a prototype online search tool which presents text and images (see SAR 2019-). The ARO is presented as a primary source edition in the original languages of Latin and Middle Scots and has not been translated into a modern language.

While we have undertaken more traditional academic investigations of this resource aimed at scholarly dissemination (e.g. Armstrong/Frankot 2020), those who have been involved in the project in various stages have expressed a shared sense of need to bring these historical primary sources to wider audiences. Project members and collaborators have directed a range of public engagement activities concentrated around raising awareness and understanding of the valuable cultural documentary heritage of these records. These archival sources speak to features of everyday life in medieval Scotland which we think are of public interest:

---

[1] The development team consisted of William Hepburn, Katharine Neil, Alana Bell and Jackson Armstrong.
[2] Including Leverhulme Trust RPG-2015-454; AHRC AH/T012854/1.

for example, the roles of women in the economy, the significance of written and spoken languages, the diversity of international trade connections, as well as how matters of social control or public health were handled by local government, and much more besides. Our objectives in making a video game were in part an attempt to advance public engagement, and especially to reach new audiences through games as a medium for history as a complement, rather than an alternative, to more conventional outputs (Bazile 2022: 856–858; Kee/Bachynski 2009: 2, 10). We considered the making of a game to be a form of *creative translation* of these primary sources analogous, in some ways, to what might be achieved by the translation of historical text into modern languages. The specific idea for a historical game about public health and disease is not new (Kee/Bachynski 2009; Zucconi et al. 2013) but we chose it as a topic likely to attract curiosity (the game was conceived before the Covid-19 pandemic but gained topical relevance in this context) as well as one which was covered in detail in the historical sources we wished to highlight and which could serve as an effective "conversation starter" on the broader subject of our research (see Bazile 2022: 866).

The project originated in 2019–2020 from two short-term grants focused on creative responses to the ARO with Hepburn as fellowship-holder and Armstrong as collaborator.[3] Those grants allowed for connections to be formed with game designer Katharine Neil, and entrepreneurship mentor Steve Aitken. In the summer of 2020 we also engaged with illustrator Alana Bell and industry mentor Michael Boniface as we developed the idea for *Strange Sickness*. After exploring various options, we settled on a crowdfunding project which ran in autumn 2020.[4] For this end we (Hepburn and Armstrong) formed a limited company, Common Profyt Games, and assembled the creative team (which consisted of us, Katharine Neil, and Alana Bell). The game was developed as a non-profit commercial project. Successful completion of the crowdfunding exercise catalysed additional support, and we embarked on development in 2021 with an approximate budget of £15,000. The game launched on Windows in late 2021 and MacOS in 2022.[5]

The institutional context in which the game idea was conceived and funded was one of public engagement with research, but when development began we revised our immediate goal. We reasoned that in order to foster any substantial public interest in our research using this approach we would need to make a game that people wanted to play. The project was in the first instance an experiment and a learning process in which the priorities of game development – creating an engaging game – were at the fore. Thus, while the project bore some resemblance

---

3   See Playing in the Archives - Exploring scope for games development with the Aberdeen burgh records (AHRC Creative Economy Engagement Fellowship: Scottish Graduate School for Arts & Humanities, 2019, https://www.sgsah.ac.uk/news/archive/headline_633508_en.html); Chivas Brothers Research Fellowship 2019-2020, https://aberdeenregisters.org/2019/07/).
4   See https://www.kickstarter.com/projects/commonprofytgames/strange-sickness.
5   See https://common-profyt-games.itch.io/strange-sickness.

to games developed as public history, such as Darren Reid's *Ab Uno Sanguine* (Reid 2024), it was not designed with specific measurements of public engagement in mind. Similarly, it was not built as an educational game with formal educational goals, even though it could be used in accordance with Jeremiah McCall's observation that a game need not have been designed for educational purposes for it to be critiqued by learners on the basis of historical evidence (McCall 2023: 22). While in some respects it aims to be a scholarly game, it does not carry "full historical arguments that can be subjected to peer review" (Clyde et al 2012: 14; Spring 2015). Neither was it intended as a serious game – a game "not primarily intended for amusement" – because the primary goal was to make a game that was engaging for players (Laamarti et al 2014: 1). We believe that we achieved this goal on the basis that the game was nominated for the BAFTA Scotland best game award and the Scottish Games Awards creativity award in 2022 and described in a feature in a leading games magazine as an "absorbing piece of interactive fiction" (Edge 2022).

That said, for the game to be an effective basis for future public engagement with research, it would also have to be reflective of our academic historical practice. Makers and scholars of historical games have often presented these two goals as oppositional (Reid 2024: 83). Moreover, the creation of historical games by academics is sometimes presented in instrumental terms, such as when historians "use the game to force the user to master concepts and rules and solve problems that are rooted in our understanding of the past" (Kee/Bachynski 2009: 3). "Fun", Kevin Kee and John Bachynski note of their own historical game design, "was not an emotion we aimed to elicit" (ibid.: 6). Our conviction in making *Strange Sickness* was that representing our academic research while also offering an engaging experience could be complementary rather than opposing goals. In addition, we approached games as a thriving art form in its own right and we wished to explore how we, as historians, could express ourselves in that form (rather than, for instance, viewing it as a popular medium to be exploited to communicate our research to the masses). To pursue these convictions, we needed to address the problem of authenticity outlined above: how could we offer an experience of authenticity that was compelling to the player, but more reflective of academic approaches to historical practice than is typically found in historical games?

Our attempt to demonstrate our convictions and address the problem of authenticity was based upon the premise of creating subject authenticity, while redrawing the parameters of object authenticity in line with current academic historical practice. This coalesced around three main pillars, to use game development terminology: one, the use of "independent style" (Juul 2019: 14) to articulate the layered relationship between game design and history; two, narrative-centric design; and three, transparent foregrounding of historical sources and interpretation. The first pillar concerns how we sought to situate the game within the wider videogames medium. The game was independent in the most technical sense of being financially independent, but also framed itself as 'aesthetically independent' (using styles and design principles that set it apart from mainstream games – e.g. hand-drawn non-realist art style) and 'culturally independent' (carrying a "cultural promise" [Juul 2019: 14] – e.g. that the game has a unique setting grounded in scholarly research on historical archives). This was a good fit with the constraints of our small budget while making a virtue of our non-conventional entry point to the field. Independent games are associated with a lower barrier for entry, which allows for a greater variety of people to make games and bring their distinctive perspectives to this art. As Jesper Juul (2019) put it, "independent games are the video games that ask us to consider how the game was made, that ask us to see them as concretely made by actual humans" (16). Following this, we sought to highlight our perspective as historians in the making of the game, as well as how the game is directly derived from historical research. Independent games and "independent style" are strongly associated with notions of authenticity (Ibid.: 12–15). So, with this approach we sought a means to elicit a subject authenticity with appeal to potential players, while also making room for a definition of historical object authenticity based on recognising the authored nature of historical representations.

The second pillar concerns the type of game we made and how it relates to the practice of history. The genre we chose was interactive fiction, a popular genre in indie game development given its relative technical low-bar and the genre's reputation for offering a wide range of subjects and perspectives. The game was prototyped in Twine and developed in Unity using the Ink narrative design plugin. Players navigated text-based dialogue, exploration and investigation over three main sections, with the narrative shaped by player choice. Scenes and characters were depicted in hand-drawn art, along with a map allowing navigation between locations. The main challenge involved identifying strategies to fight the plague and assembling supporting evidence for those strategies to convince fellow councillors in the story. The strategies chosen affected the narrative in terms of the reactions of different townspeople. Interactive fiction is a genre whose suitability for historical games is highlighted by Kee et al. (2009). In the interpretation of sources, historians construct narratives to make sense of the past and this can function on various levels: for example, the sequence of events in a local uprising, trends in the use of written documents over time, or changes in medical or legal knowledge spread through scholarly networks – but all take the form of narratives built by historians. What is more, the matter of the boundary between historical and fictional narrative is not always clear (Trouillot 1995: 8; Munslow 1997: 36–119; Munslow 2007; Jordanova 2019). For creative writing (in which writing for narrative-driven games and interactive fiction can be situated), the genre of historical fiction wrestles with the same issue, but from the opposite side. From this perspective Hilary Mantel (2019) reflects on her craft, writing that "[f]iction, if well written, doesn't betray history, but opens up its essential nature to inspection" (19). Clyde et al. (2012) have even suggested that "narrative is an important factor in making the gamic mode of history recognizable as history" (8). By centring narrative writing in *Strange Sickness*, we sought to emphasise history as a contingent, authored practice and to offer an appealing authenticity in line with indie aesthetics.

The third pillar, which we explore in detail in the remainder of the paper, is the bedrock of how the game presents its relationship to historical source material. This has been a key concern for other historians engaged in game development (e.g. Clyde et. al. 2012: 5). Whereas the image of the historian revealing objective facts from documents which speak for themselves is now essentially discarded, what has replaced this is an emphasis on the key role of historian as a selector and analyst of primary sources (the traces of the past), wielding a creative voice in that process (Skinner 1997). Late twentieth-century postmodernism, which informed the linguistic turn of the professional historical discipline and challenged understandings of "truth" and "knowledge" (Surkis 2012; Evans 1997), have also had their impacts, but decades on from this historical practice has not discarded their reliance upon primary sources that require interpretation (Carr/Lipscomb 2022; Cannadine 2002; Carr 1961). As well as supporting our goal of showcasing research in which the transcription of medieval sources was a key component, we aimed for transparency about how we used our sources. With these three pillars, we endeavoured to emphasise the role of the historian in shaping narrative from those primary sources as a means of framing the game's approach to history, thereby stimulating subject authenticity that would appeal to players through an approach to object authenticity consistent with our broader academic practice.

Reflections on sources and referencing in *Strange Sickness* and other games

In the remainder of this article, we reflect on how we centred references to historical research and sources in *Strange Sickness* as the key part of our development process, with the intention of addressing the problem of authenticity. We also reflect on other games to draw broader conclusions about approaches to academic-led development of historical games, always bearing in mind that these other projects were shaped with different ambitions and constraints. We discuss the use of primary sources from the point of view of developers making design choices, but in making those choices, as will be seen, we take into consideration the distinctive quality of games as a medium in that players at least to some extent shape the representation they encounter, and thus function as "player-historians" (Chapman 2016: 22). An important consideration here is the task of *creative translation* that drove the *Strange Sickness* project and other creative efforts to bring research on Aberdeen's medieval records to new audiences – the idea that the sources contain much of public interest that is largely inaccessible in its original form in Latin and Scots texts. These documentary sources can remain obtuse to the non-specialist reader even when translated into modern English, given the formulaic nature of legal and administrative records. Thus, unlike some historian-developed games such as *Ab Uno Sanguine* (Reid 2024: 87), use of document texts in original or lightly-adapted form was an approach we avoided.

However, the text-based nature of *Strange Sickness* provided the potential for referencing analogous to that used in academic publications. A game that tends in this direction is *Walden, A Game* (USC Game Innovation Lab 2017), described by Dawn Spring (2015) as offering "a unique example of playing a historical moment" (210). It is a videogame adaptation of transcendentalist writer Henry David Thoreau's *Walden*, an influential account of his time living by Walden Pond in New England. The game broadly fits in the survival genre, and the player is tasked with finding food, building shelter and performing chores for neighbours to earn money. As players carry out these tasks and explore the pond's environs, they populate their journal with entries. In practice this means that encountering things in the game world triggers fragments of text from *Walden*, which are then assembled (in the order the player encounters them) in the player-as-Thoreau's in-game journal. These are not technically cited but can be easily found within online texts of *Walden*. While an approach like this clearly makes more sense when using a single, well-known text – and one that lends itself very well to the survival genre – it nonetheless provides a model for how comprehensive referencing might be used even in a game in which the primary mechanics are non-textual.

The ambition for *Strange Sickness* was that every piece of historical information used in the text of the game would be traceable to its root source, with an explanation of how the historian-developer had adapted this information for use in the game. Such an approach is advocated by Clyde et al. (2012) who argued that "a gamic mode of history needs to be particularly rigorous on the points of evidence and interpretation" (6) and highlighted the suitability of interactive fiction for doing so. In practice, this would have meant in some cases just a citation for something used exactly as it appears in the source, some commentary when the historical information required minor adaptation to fit the context of the game, and longer explanations for pieces of information deemed necessary for the game though not furnished directly by the sources but rather by inference from the ARO and comparable records. Given that even getting the basic core of the game functioning was a stretch with our small budget, it quickly became apparent that this task would be impossible under project constraints, and we sought other ways to achieve our aims.

The most common way in which historical games make reference to past reality is not through source references but through an in-game encyclopaedia or glossary (Houghton 2024: 58). These explain the settings, events, people and other aspects of the past on which the game is based. They are the historical game's equivalent of encyclopaedias used in games more generally, for example in a fantasy RPG such as *The Witcher 3* (CD Projekt Red 2015) to explain the world or universe in which the game is set, describing characters, places historical events in that world amongst other subjects – everything comprising the invented world's "lore" (Anderson 2019: 181). Describing a fictional world (albeit in many cases, including *The Witcher 3*, based on pre-existing intellectual property) with omniscient knowledge in these cases makes sense, as the game makers are here describing a world which they have conjured. This is communicated to the player directly, outside of the diegetic world of the game.

This approach is followed in some historical games, with the difference that the world described, often from what reads as a similarly omniscient standpoint with perfect knowledge, is the real world and not the creation of the game's developers. One of the most-discussed game series in historical game studies, *Sid Meier's Civilization*, provides a well-known example. These games present a *pot pourri* of fragments about the past – famous people, places, buildings and so on – which the player can endlessly remix. From the beginning these games have included a 'civilopedia', which serves as a reference to game mechanics and background on the subjects covered, thus presenting itself as a factual underpinning to the remixing of history encouraged by the gameplay.

The civilopedia in *Civilization VI* (Firaxis Games 2016) makes incidental reference to historical texts featured in the game such as Niccolò Machiavelli's *The Prince* (2011 [1532]). In places it also references the messy business of scholarship on which our understanding of the past is based, such as in its entry on cartography when it acknowledges "a fair amount of scholarly debate" around the subject of early maps. For the most part though discussion of the past follows the style of the entry on the printing press:

"The invention revolutionized the world and gave rise to mass communication. It spread rapidly across Europe, since booksellers could now make (and sell) lots of copies. It not only standardized language and knowledge – bringing page numbers, tables of contents, indexes, the ability to cite other works and all sorts of things not possible with hand-copied books – but it taught humanity to think in linear (since that's how folk read now) terms rather than holistically. And it sparked – or at least promoted – the Scientific Revolution and the Reformation. So, if civilization is a mess, blame it on Gutenberg" (Firaxis Games 2016: n.p.).

There is nothing here that is likely to be contradicted by scholarly consensus but nevertheless sweeping claims are made with little or no qualification and no references are provided. *Civilization VI*, and other examples such as *Kingdom Come: Deliverance* (Inderwildi 2018: Schwarz/Weber 2023), refer to the history behind their games but not, except incidentally, to their sources. Their encyclopaedias, as well as functioning to help players navigate the game, make claims to authenticity largely without reference to primary sources or the role of historians in shaping knowledge of the past, presenting this historical source material as a more or less pristine bedrock of fact without, or with very few, references to sources or the process of historical interpretation. In such passages the sources of history and the process of building an understanding of the past from them are largely obscured within a varnished black box of historical description.

Examples of the use of citations of primary sources within games, in the context of in-game encyclopaedias, can be found. *Assassin's Creed Mirage* (Ubisoft 2023) offers a good recent illustration of this. *Mirage's* presentation of the history underpinning the game is broadly in line with other big budget games that use encyclopaedias, like *Civilization* and *Kingdom Come: Deliverance*. The frame in which encyclopaedic information in *Assassin's Creed* is presented is also somewhat more involved on account of the premise of the franchise that the player controls a character in the present-day who accesses the past through a science fiction technology that enables them to relive the genetically-stored memories of their ancestors (though it should be noted that the text of the entries is non-diegetic even within this frame, referring as it sometimes does to choices made in the development of the game). This framing is much less pronounced in more recent *Assassin's Creed* games such as *Mirage*, but there are still traces of it and it offers a diegetic explanation of why vignettes about the history behind the game can be accessed through floating spheres found in the game world. Unlike most such encyclopaedias, *Mirage* includes fully-referenced primary sources. Most of these are museum pieces or visual sources, but some documents are included. Sometimes they are a direct reference point for something in the game world, such as an architectural detail, and in other cases they are more broadly illustrative of the historical topic encountered in the encyclopedia entry. Overall, the primary sources presented have a loose relationship to the game's core text compared to, say, primary sources referenced in historical academic publications. Nevertheless, their presence provides a pathway players can use to identify particular objects in museum collections. Even if a player does not visit a physical museum cited in this way or look for information about the objects online, as most will not, sources and their citations point to the authenticity of the game as something constructed from historical sources rather than the flat, positivistic presentation of historical fact found in games such as *Kingdom Come: Deliverance*.

A key goal in the development of *Strange Sickness* was to foreground the historical sources and research on which the game was based in a manner consonant with modern academic historical practice which emphasised, rather than hid, that the game's account of history was authored. This was part of our efforts towards what Julien Bazile terms "intellectual honesty" – that is, to "discourage unfruitful confusion between the game as a historical narrative and the game as an objective, exhaustive recreation of the past itself" (Bazile 2022: 65). As such, we were keen to avoid presenting history as unproblematised fact like the positivistic encyclopaedias of games such as *Civilization VI*. Still, we recognised the benefits of including a glossary in the game in providing a tool for the player to learn the definition of terms, contextualise the game's narrative and to ensure the game's script avoided heavy exposition and repetition. The approach we took is similar to that taken in the encyclopaedias of fantasy games such as *Avowed* (Obsidian Entertainment

2025), in which the glossary is intended to reflect the knowledge of the player-character about the in-game world in order to avoid the need for expositionary dialogue. In *Strange Sickness*, the glossary was presented as pieces of the player-character's internal monologue on terms and subjects familiar to the character, but not necessarily to the player. This allowed the glossary to be a player aid (with a collecting mechanic as terms were encountered in the story) as well as a canvas for further character development. At the same time, this approach makes it clear that the glossary entries are the player-character's subjective assumptions about his world, rather than an implied claim of perfect knowledge of the past. While in practice this information is based on evidence or best guesses by historians making the game, it is presented as part of the game's diegesis rather than as a claim to external historical truth.

This approach to the glossary within *Strange Sickness* sought to escape the trap of a positivistic approach to the game's underlying historical sources but, given the diegetic nature of the glossary, does not provide an opportunity for referencing. However, there are other opportunities in the development of historical games for showing the link between the game and its sources in ways that met our key ambitions. Some games foreground the role of primary sources in the creation of history through design choices highlighting the form and content of primary sources as a fundamental part of the game's aesthetic, story or mechanics without necessarily referring directly to actual primary sources. Two recent examples, *Inkulinati* (Yaza Games 2023) and *Pentiment*, have been heavily influenced in their design by the documentary sources of the medieval and early modern periods. Medieval manuscripts were the key inspiration for *Inkulinati*, and art director Dorota Halicka noted that the "idea of enclosing everything in a book was a very important decision that strongly influenced the whole thing" (in Kerr 2021: n. p.), from the gameplay mechanics to the interface. In a more explicit manner, Darren Reid's *Ab Uno Sanguine* includes art assets which were created as copies of historical documents (Reid 2024: 98).

Pentiment takes this approach to historical primary sources even further. While the game's narrative and characters are fictional, it draws closely on historical source material from Germany in the early sixteenth century, rooting its story in a highly specific frame of historical reference. Again, these sources are not directly referenced in game, except in parts of the script that diegetically refer to real historical texts from the period such as those found in the library which features in the game's monastic setting. However, the game's core design is shaped by, and directly invokes, primary historical sources of the period. The player-character, and the story's protagonist, is a manuscript illuminator. The game's text is rendered in an approximation of contemporary fonts with animations that reveal the text as if it is being written on the screen, complete with quill scratching sound effects, with mistakes sometimes made and scored out then replaced with the corrections. Similar to *Inkulinati*, parts of the game's user interface (UI) are presented as pages of a book. The game's plot and core mechanics of solving a murder mystery have no set solution, reflecting, according to the game's director Josh Sawyer, the ambiguity of historical truth as revealed by primary sources. The player, through their choices, essentially has to decide the narrative that will stand as the truth, just as a historian attempts to make order from the fragmentary window onto the past offered by surviving sources. As Esther Wright has noted, "Pentiment is remarkable because of its deliberate reflexivity on the nature of history as a process and practice and of the writing of history as layers built up over time by different people with competing perspectives and motivations" (Wright 2024: 191).

Thus in both cases historical sources are not referenced in-game in a conventional sense, but the game's core design choices reference, and bring the player's attention to, a genre of historical source material. They also point to an understanding that history is our partial present knowledge of the past built on mutable interpretation of surviving evidence, rather than something which can be fully recovered, recreated and represented. In *Strange Sickness*, while the form of records was not as central a theme as in these two games, it was still an important influence on the design. The in-game map is presented diegetically as a map on a table (based on a historical map surviving from a later period), while other elements of the UI visualise aspects of record-making and keeping. The process of information-recording is also referenced in a conceit whereby the town clerk assists the player-character by taking notes of his investigation in his notebook. This is also the notebook in which he drafts entries for Aberdeen's municipal council registers (i.e. the real historical records on which the game is largely based). This functions both as a log for the player to track progress as well as a form of incidental worldbuilding through adapted quotations from the historical records, interspersed between log entries and distinguished by italics. Furthermore, as project developer Katharine Neil noted, the core mechanic of the game – assembling evidence and using it to present an argument – was intended to reflect the practice of historians, and this was reinforced in the game's narrative by the player-character's ambition to be a scholar (Edge 2022).

Another means by which games can reference primary historical sources is through what can be called their paratexts – that is to say, texts or artifacts surrounding a primary text (Consalvo 2017: 177). In the case of games, this might include box art, game guides and marketing materials, and this has proved a rich line of enquiry for scholars (Wright 2023; Wright 2018). Here we are interested in paratexts offered by the game's creators which provide some explanation of a game's relationship to historical primary source material. An example is *Echoes of History*, a podcast series by Ubisoft providing historical background on their historical games, primarily the *Assassin's Creed* series (Assassin's Creed: Echoes of History 2022). *Mirage*, discussed above, was very well-contextualised by this and other means, including interviews with academic consultant Glaire Anderson, emphasising authenticity through the game's roots in research while acting as a platform for creative freedom to showcase Islamic art more broadly (in Walton

2023). Historical games with paratexts referencing sources can also be found at the other end of the scale of game-making from big budget games. The Historically-Accurate Game Jam has run for the last several years, and the games submitted as part of it are available via itch.io. Looking at the collected submissions through the different iterations of the jam reveals how the submission process has developed to incorporate statements from developers about the historical basis of the game, with "historical accuracy" becoming one of the criteria by which submissions were judged. Some of these statements include discussion of primary sources such as the statement of @JBOStudios about their game *A Local Revolt; The Luddites* (JBO Studios 2023) that "[t]he resources I used included websites and newspaper archives. Visits to the local area, museums, and historical buildings I also spoke with the locals and some rather elderly people in order to get a feel for the story" (Historically Accurate Game Jam 7 2023).

In the development of *Strange Sickness*, the use of paratext emerged as a key strand of communicating the game's sources and underpinning research. This was built into the project from the beginning in the materials created for the game's crowdfunding campaign and surrounding blogs, social media posts and media coverage, which was reiterated and refined through further coverage around the release of the game (see Edge 2022). However, the key paratext of *Strange Sickness* is a website (strangesickness.com) built to accompany the game, which includes a 'historians' commentary' that is linked from the game's title page and credits. This offers short written pieces on how key aspects of the game were adapted from sources and links for players who wish to go further and consult the source material via the wider Aberdeen Burgh Records research project sites, other online sources, and secondary publications on relevant topics. For instance, it explains that the game's map is based on a map from 1661, some 150 years after the period of the game, because there was no earlier equivalent, and explains how this relates to the town as it would have appeared in ca. 1500 and the choices and compromises made in the adaptation of the map. The commentaries also offer a statement on the game's theory of history: "that the past cannot be reconstructed and experienced exactly as it was, but only interpreted through fragmentary survival of historical evidence" (Common Profyt Games 2021, n. p.).

To summarise, while a comprehensively referenced game might be possible, budget constraints made it unfeasible in *Strange Sickness*. The most common means of referring to historical source material in games is through in-game glossaries. These often serve to reinforce the positivistic assumptions underlying historical games. While avoiding this and serving a gameplay function, the diegetic glossary in *Strange Sickness* could not be used to refer to source material and the historical research process. As other games like *Pentiment* demonstrate, game elements can be used to point to the sources and process of history in a way that is sufficient to signal to players that our understanding of the past is built on surviving primary source evidence and the influence of the historians working with this evidence. As such, references were not needed to make this essential point, which was instead achieved through such means as visual elements referring to source materials, the historical research-adjacent key mechanic, and the conceit of the player's log being the town clerk's notebook. Finally, while this offered a headline message about the game's approach to history, we acknowledged the role of the player in shaping their experience and that players may wish to engage with the game's source material and themes beyond this level, echoing the efforts of other historian-developers such as Darren Reid to offer different levels of depth at which players can engage with history though the game (Reid 2024: 86–87). This was offered through paratext, chiefly the supporting website linked from the game itself.

Conclusion

In creating a historical game, any historian-developer, along with the particular demands of their project, is faced with the question of how to address the problem of authenticity. In short, this is the dilemma presented by the historical game player's demand for authenticity on the one hand (usually understood in positivistic terms) and on the other the academic profession's demand for rigour and credibility in which the same positivism is outmoded and insufficient. Through reflection on the development of *Strange Sickness*, we argue that these demands can, however, work to the same end: that the need for academic credibility and an appealing (i.e. authentic) game, as well as the constraints of low budget, can all converge to serve as a fruitful creative force. This requires us to find ways for the game to highlight underlying sources and the process by which they have been interpreted by the game makers, thus adopting the aesthetics of independent games in which a sense of authorship is a key attraction. This can advertise to players that authenticity lies not in any claims to replicate a past reality but in an effort to portray the practice of modern-day historians. The result is to extend an invitation to players to engage in this practice themselves as they encounter representations of the past, to a level of their choosing.

References

"Aberdeen Registers" (2016). Retrieved from https://aberdeenregisters.org/

Anderson, S. L. (2019). The Interactive Museum: Video Games as History Lessons through Lore and Affective Design. E-Learning and Digital Media 16(3), pp. 177-195.

Anthropy, A. (2012): Rise of the Videogame Zinesters. New York: Seven Stories Press.

Armstrong, J.W./Frankot, E. (eds) (2020): Cultures of Law in Urban Northern Europe: Scotland and its Neighbours c.1350–c.1650, London: Routledge.

ARO (2019): "Aberdeen Registers Online: 1398-1511." Retrieved from https://www.abdn.ac.uk/aro.
"Assassin's Creed: Echoes of History" (2022). Retrieved from https://www.ubisoft.com/en-us/game/assassins-creed/news/5SeepXEMoLEUbIK6UImQNF/assassins-creed-echoes-of-history.
Bazile, J. A. (2022): ""An 'Alternative to the Pen'? Perspectives for the Design of Historiographical Videogames." Games and Culture 17(6), pp. 855–870.
Bloch, M. (1992 [1954]): The Historian's Craft. Manchester: Manchester University Press.
Cannadine, D. (ed.) (2002): What Is History Now?. Basingstoke: Palgrave Macmillan.
Carr, E. H. (1961): What is History? London: Macmillan.
Carr, H./Lipscomb, S. (2022): What is History, Now? How the Past and Present Speak to Each Other. London: Weidenfeld & Nicolson.
CD Projekt Red (2015): The Witcher 3: Wild Hunt [Video Game]. CD Projekt.
Chapman, A. (2016): Digital Games as History: How Videogames Represent the Past and Offer Access to Historical Practice. London: Routledge.
Chapman, A./Foka, A./Westin, J. (2016): "Introduction: What Is Historical Game Studies?" Rethinking History 21(3), pp. 358–371.
Clyde, J./Hopkins, H./Wilkinson, G. (2012): "Beyond the 'Historical' Simulation: Using Theories of History to Inform Scholarly Game Design." Loading… 6(9), pp. 3–16.
Common Profyt Games (2021): Strange Sickness [Video Game]. Common Profyt Games.
Consalvo, M. (2017): "When Paratexts Become Texts: De-centering the game-as-text." Critical Studies in Media Communication 34 (2), pp. 177–183.
Cook, J. (2024): "Kingdom Come: Deliverance and the aesthetics of authenticity." Journal of Sound and Music in Games 5(2), pp. 23–48.
Cruz Martínez, M. A. (2020): The Potential of Video Games for Exploring Deconstructionist History. PhD Thesis: University of Sussex.
DeVine, D. (2022): "Declaiming Dragons: Empathy Learning and The Elder Scrolls in Teaching Medieval Rhetorical Schemes." In: R. Houghton (ed.), Teaching the Middle Ages through Modern Games. Berlin/Boston: De Gruyter Oldenbourg, pp. 69–86.
Edge (2022): "A plague tale: The story about a historical game about an eerily familiar pandemic." In: Edge, March 2022, pp. 14–15.
Evans, R.J. (1997): In Defence of History. London: Granta Books.
Firaxis Games (2016): Civilization VI [Video Game]. 2K Games.
Gundermann, C. et al. (2025): "Authenticity." In: C. Gundermann et al. (eds), Key Terms of Public History. Berlin/Boston: De Gruyter Oldenbourg, pp. 9–32.
Hepburn, W./Small, G. (2020): "Common Books in Aberdeen, c. 1398–c. 1511." In: J. W. Armstrong/E. Frankot (eds), Cultures of Law in Urban Northern Europe: Scotland and its Neighbours c.1350–c.1650. London: Routledge, pp. 41–57.

Historically Accurate Game Jam 7 (2023). Retrieved from https://itch.io/jam/historically-accurate-game-jam-7/rate/2195465.

Houghton, R. (2024): The Middle Ages in Computer Games: Ludic Approaches to the Medieval and Medievalism. Cambridge: D. S. Brewer.

Inderwildi, A. (2018): "Kingdom Come Deliverance's Quest for Historical Accuracy Is a Fool's Errand" (5 March 2018). Retrieved from https://www.rockpapershotgun.com/kingdom-come-deliverance-historical-accuracy.

JBO Studios (2023): A Local Revolt The Luddites [Video Game]. JBO Studios.

Jordanova, L. (2019): History in Practice. 3rd edn. London: Bloomsbury Academic.

Juul, J. (2019): Handmade Pixels: Independent Video Games and the Quest for Authenticity. Cambridge, MA: MIT Press.

Kee, K./Bachynski, J. (2009): "Outbreak: Lessons Learned from Developing a 'History Game'." Loading ..., 3(4), pp. 1–14.

Kee, K./et al. (2009): "Towards a Theory of Good History Through Gaming." Canadian Historical Review 90(2), pp. 303–326.

Kerr, C. (2021): "Thrills and Quills: The Ink-Based Combat and Magical Manuscripts of *Inkulinati*" (22 November 2021). Retrieved from https://www.gamedeveloper.com/design/thrills-and-quills-the-ink-based-combat-and-magical-manuscripts-of-inkulinati.

Kleinerman, D./Haynes, C. (2023): Regret in Play and in Paint: Authorship, Narrative, and Intertextuality in *Pentiment* (2022). Proceedings of DiGRA 21, n. p.

Laamarti, F./Eid, M./El Saddik, A. (2014): "An Overview of Serious Games" International Journal of Computer Games Technology, 2014(1), pp. 1–15.

Machiavelli, N. (2011 [1532]): The Prince. Translated by T. Parks. London: Penguin Classics.

Mantel, H. (2019): "Resurrection: The Art and Craft." In: Remarkable Minds: A Celebration of the Reith Lectures. London: BBC Books.

McCall, J.B. (2023): Gaming the Past: Using Video Games to Teach Secondary History. 2nd edn. New York: Routledge.

MicroProse (1991): Sid Meier's Civilization [Video Game]. MicroProse.

Munslow, A. (1997): Deconstructing History. London: Routledge.

Munslow, A. (2007): Narrative and History. Abingdon: Routledge.

Obsidian Entertainment (2022): Pentiment [Video Game]. Xbox Game Studios.

Obsidian Entertainment (2025): Avowed [Video Game]. Xbox Game Studios.

Partner, N. (2012): "Foundations: Theoretical Frameworks for Knowledge of the Past" In: Partner, N./Foot, S. (eds) (2012): The SAGE Handbook of Historical Theory. London: SAGE Publications, pp. 1–8.

Reid, D. (2024): Ab Uno Sanguine [Video Game].

Reid, D. (2024): "Video Game Development as Public History." The Public Historian 46(1), pp. 74–107.

SAR (2019-) "Search Aberdeen Registers." Retrieved from https://sar.abdn.ac.uk/.

Schwarz, A./Weber, M. (2023): "New Perspectives on Old Pasts?" Arts 12(2), pp. 69–89.

Skinner, Q. (1997): "Sir Geoffrey Elton and the Practice of History." Transactions of the Royal Historical Society 7, pp. 301–316.

Spring, D. (2015): "Gaming History: Computer and Video Games as Historical Scholarship." Rethinking History 19(2), pp. 207–221.

"Strange Sickness" (2021). Game website. Retrieved from https://strangesickness.com

Surkis, J. (2012): "When Was the Linguistic Turn? A Genealogy." American Historical Review 117(3), pp. 700–722.

Trouillot, M.-R. (1995): Silencing the Past: Power and the Production of History. Boston: Beacon Press.

Walton, W. (2023): "Rebuilding Baghdad – in the New Instalment of Assassin's Creed" (11 October 2023). Retrieved from https://www.apollo-magazine.com/assassins-creed-mirage-baghdad-video-game-islamic-art-architecture/.

Ubisoft (2007-2025): Assassin's Creed [Video Game series]. Ubisoft.

Ubisoft Bordeaux (2023): Assassin's Creed Mirage [Video Game]. Ubisoft.

UNESCO (n. d.): "Memory of the World". Retrieved from https://unesco.org.uk/our-sites/memory-of-the-world.

USC Game Innovation Lab (2017): Walden, A Game [Video Game]. USC Games.

Warhorse Studios (2018): Kingdom Come: Deliverance [Video Game]. Deep Silver.

White, H. (1987): The Content of the Form: Narrative Discourse and Historical Representation. Baltimore: Johns Hopkins University Press.

Wright, E. (2018): "Promotional Context of Historical Video Games." Rethinking History 22(4), pp. 598–608.

Wright, E. (2022): "Still Playing With the Past: History, Historians, and Digital Games." History & Theory 61(4), pp. 166-177.

Wright, E. (2023): "Paratexts, 'Authenticity,' and the Margins of Digital (Game) History." In: R. Seiwald/E. Vollans (eds), (Not) In the Game. Berlin: De Gruyter, pp. 33–56.

Wright, E. (2024): "Layers of History." ROMchip: A Journal of Game Histories 6(1). Retrieved from https://romchip.org/index.php/romchip-journal/article/view/191

Yaza Games (2023): Inkulinati [Video Game]. Daedalic Entertainment.

Zucconi, L./Watrall, E./Ueno, H./Rosner, L. (2013): "Pox and the City." In: J. Dougherty /K. Nawrotzki (eds), Writing History in the Digital Age. Ann Arbor: University of Michigan Press, pp. 198–206.

Historians Making Games
Unveiling a History Game Design Ethos

*Magnus Henrik Sandberg, Eirik Brazier,
and Ragnhild Hutchison*

Abstract

This article examines the design ethos of a historical video game developer and one of its games. The study investigates how the company which was founded and run by trained historians navigates tensions between historical authenticity and compelling gameplay. Drawing on Theodor Adorno's concept of mimesis and Jörn Rüsen's theory of historical consciousness, the research employs qualitative analysis of game design principles and developer motivations. The analysis reveals that these developers position themselves as disseminators of historical research, emphasising commitment to historical sources and resistance to oversimplification for dramatic effect. Based on this case study, the authors propose the concept of critical authenticity as a theoretical construct for discussing how historical games can maintain scholarly rigour while addressing the challenges of commercial pressures and player expectations in the gaming medium.

Keywords

design ethos, historical authenticity, critical authenticity, historical consciousness, mimesis, historian-developer

1. Introduction

The Widow's Boutique (2023) takes players to 1820s Christiania (present-day Oslo), where they assume the identity of a widowed drapist striving to maintain her husband's business. The game invites players to explore both the challenges and opportunities that shaped women's societal and economic roles while also illuminating aspects of early capitalist development in Norway and the broader context of Western Europe during the early 1800s.

The game constructs its historical setting through the integration of primary source materials, such as contemporary newspaper articles, daily temperature records, authentic fashion and furniture designs, and real historical figures and locations. Most compelling, however, is how Tidvis, an Oslo-based studio that approaches game development from a historian's perspective, has meticulously crafted the game's narrative and plot to critically disseminate the primary sources used. According to the studio, their mission centres on making historical research accessible to a broad audience through interactive media (Tidvis n.d.a). Hence, the game presents itself as both a narrative experience and a carefully researched historical document, demonstrating how interactive media can effectively convey historical research to a modern audience. This dual identity – as both narrative and historical documentation – raises important questions about the values and assumptions that inform the development of historically grounded games. To understand how such games come into being, this article looks beyond the game's surface-level content and examines the ethos guiding the creators.

Historians, as well as game designers, work according to a more or less tacit professional ethos, understood as the fundamental character and beliefs that shape the profession's identity and culture. It seems reasonable to assume that the professional ethos of historians at work differs from that of game designers. Yet, game designers often tell stories with explicit reference to historical periods, incidents, and personae. One may wonder if the underlying values of such design processes differ from the process of designing games of pure fiction. This article asks the question – what constitutes the *design ethos* of historical game-making? To explore this, the article focuses on a single case: the Oslo-based studio Tidvis and its game *The Widow's Boutique*. Tidvis is explicitly committed to combining rigorous historical research with interactive storytelling. By examining both the studio's stated intentions and the design choices evident in the game itself, the aim is to shed light on how a historically informed design ethos takes shape in practice – how historical fidelity, narrative engagement, and educational ambition are negotiated in the development of a game.

2. Research review

Recent scholarship contends that video games constitute a medium that both widens the repertoire of how the past may be represented and introduces distinctive methodological and epistemological challenges for historians. In response, the field of *historical game studies* has taken shape, shifting discussion away from narrow verdicts on factual accuracy to examine "broader and more productive questions concerning the nature, possibilities, uses and limitations of the game form itself" (Chapman et al. 2017: 360). In line with postmodernist and poststructuralist historical theory (cf. Southgate 2005), historical games have been described as *resonating* with history rather than (accurately) representing it. Chapman (2018), paraphrasing Apperley (2010), describes resonance in historical games as "the sensation of interpreting the representation of the game as relating to something other than only the game's rules, as referring to something not entirely contained within the game" (Chapman 2018: 35). Thus, 'real' history is there, but rather than seen as accurately represented by what Chapman calls the *developer-historian*, it is also a product of both the subjective *reading* and the constructive *doing* of the player (Chapman 2018).

Copplestone's (2017) empirical research reveals significant tensions in how different stakeholders conceptualise accuracy in cultural-heritage videogames, with developers, players, and heritage practitioners often operating from fundamentally different epistemological frameworks. Her findings demonstrate that while traditional notions of accuracy focus on fidelity to source materials, the interactive nature of games introduces new questions about authenticity that encompass experiential and systemic accuracy. Authenticity may be a more versatile term, often strongly overlapping with accuracy in reconstructivist contexts and among historians and other 'cultural-heritage professionals', while developers and players tend to understand authenticity more as believability. Related to the latter, Salvati and Bullinger (2013) discuss a *selective authenticity* where, in WW2 games, accurately represented weapons and war machinery build authenticity among the player community.

In these matters, historical games are, however, not necessarily that different from other kinds of games, nor for that matter, other forms of history dissemination. In game studies, Ian Bogost's concept of *procedural rhetoric* posits that games convey meaning through their mechanics and rules, thereby offering a unique form of argumentation or expression (Bogost 2007). As argued both by Clyde et al. (2012) and by Kapell and Elliott (2013) in the introduction to their edited collection, given that historical models operate on similar principles of rule-based systems, the use of game mechanics to simulate historical processes presents a compelling method for educational and analytical purposes that may be not that different from the historian's work process.

History is, however, also told in visual, auditive, and verbal modalities, and most games are designed as complexes of several of these. For this study, it is of particular interest that *visual novels*, where the aesthetic presentation is as essential as the interactivity, have been recognised as a suitable game genre for disseminating historical insights (Øygardslia et al. 2021). A visual novel is an interactive digital narrative game that combines textual storytelling, static or animated visuals, and player-driven decision-making to create an immersive gaming experience (Cavallaro 2010). *The Widow's Boutique* is, as we shall see, an expansion of the traditional visual novel structure, incorporating elements of strategy, specifically a business simulation mechanic that also considers social status and civic orientation (Lebowitz/Klug 2011).

Finally, historical game research has explored the conversations surrounding the game and found that such arenas offer different opportunities for increased learning, deeper understanding, and critical perspectives (Rahimi et al. 2020). Such conversations can also be understood as *paratexts*, with the potential to both situate and expand on the game proper and even to take over as the main text to which the game will relate (Genette 1997; Consalvo 2017). Connected to this is also a critical focus on game development that takes into account the broader picture, encompassing not only what the games communicate through their rhetoric but also how they are created. Hammar (2023), for instance, argues that the economic structures, labour conditions, and corporate cultures of game studios like Rockstar directly influence the kinds of histories and representations that appear in games such as Red Dead Redemption 2. By analysing how factors such as profit motives, workplace hierarchies, and global divisions of labour shape creative decisions, Hammar demonstrates that the way games are made is inseparable from the stories they tell and the perspectives they reinforce. In a complementary vein, Bódi (2022) explores how a game's design ethos, shaped by both industry practices and developers' intentions, frames the possibilities for player agency and meaning-making. Bódi's approach, which has inspired the methodological approach of this study, connects paratextual materials and design discourse to production, showing that the affordances and constraints of a game are deeply rooted in the context of its development. Together, the works of Hammar and Bódi highlight that to understand games critically, we must situate them within their broader production environments, recognising how industry structures, design philosophies, and material conditions shape both form and content.

While maintaining a dialogue with previous research findings in the field of historical and general games research, we will build this study on a theoretical foundation of aesthetics and historical consciousness.

3. Theory

To critically examine the design ethos of historical game-making, it is necessary to engage with theoretical frameworks that explore how history is both represented and experienced in games. In this section, we draw on two complementary perspectives: Theodor Adorno's concept of *mimesis*, which offers insight into the aesthetic and ideological dimensions of historical representation, and Jörn Rüsen's theory of *historical consciousness*, which highlights the narrative and ethical implications of engaging with the past. Together, these approaches help to analyse how historical games like *The Widow's Boutique* navigate the tension between authenticity, storytelling, and market-driven design.

Adorno's concept of mimesis (Adorno 2014 [1970]; Weiss 2011) offers a valuable lens for understanding the aesthetic dimension of historical video games, such as *The Widow's Boutique*. Mimesis, in Adorno's view, is not merely a reproduction of reality but an active, dialectical process of engaging with it. In the context of historical games, mimesis manifests in the effort to recreate the 'feel' or essence of historical periods through visual design, narrative, and gameplay mechanics. However, Adorno's critique of the culture industry warns against the commodification of art, which can strip it of its critical potential. Hence, historically accurate video games, as products of the entertainment industry, risk reducing complex historical realities to consumable spectacles. An example of such commodification can be seen in what Salvati and Bullinger (2013) call 'selective authenticity', where, for instance, technical war machinery used in World War II is meticulously recreated in a game, while a traditional narrative about heroes and villains might be passed down uncritically. Thus, Adorno highlights how the tension between mimesis as creative engagement and the demand of commercial viability poses critical questions for the design ethos of historical games: to what extent does the game's aesthetic design prioritise historical authenticity over marketability, and how does the mimetic process in historical games balance fidelity to historical sources with the need to create an engaging player experience?

Jörn Rüsen's theory of historical consciousness complements Adorno's aesthetic critique by emphasising the ethical and narrative dimensions of engaging with the past (Rüsen 2004). Rüsen argues that historical narratives serve as tools for understanding temporal change and guiding moral action. In this sense, historical video games are, just like other cultural objects, not merely vehicles for entertainment but also media through which players encounter and interpret the past. The design ethos of historical game-making can be critically evaluated through the lens of Rüsen's concept of multiperspectivity, which emphasises the inclusion of diverse historical voices and perspectives. Key questions include whether the game's narrative fosters critical engagement with history or perpetuates simplistic or one-dimensional interpretations, how marginalised or non-dominant perspectives are represented within the game's historical framework, and to what extent the game challenges players to grapple with the moral and ethical complexities of historical events.

To capture the particularities of historical game design we apply Bettina Bódi's concept of design ethos (Bódi 2022: 3). In this context, ethos refers to the core values and guiding principles that shape the attitudes and decisions of a person, group, or organisation. Bódi has suggested that the design ethos of a game development studio can be reconstructed by studying how the studio communicates its professional and artistic identity publicly, in the trade press, as well as in more general media outlets. This material, together with other paratexts such as game reviews and journalistic coverage of the studio and its games, as well as the aesthetics of the games themselves, constitutes Bódi's building blocks for reconstructing a game developer's design ethos. The process may be challenging due to what Bódi refers to as an "invisible wall" of secrecy typical for the industry (Bódi 2023: 3).

Unlike AAA game studios that often maintain high levels of secrecy, historical scholarship operates with greater transparency. Codifications of the professional values of historians tend to emphasise open critical discourse, clear methodological practices, fair and empathetic representation of historical subjects, and collaborative knowledge building (cf. Statement on Good Practice for Historians n.d.; Statement on Standards of Professional Conduct 2023). This transparency in historical practice provides better insight into how history-focused game developers approach their work compared to mainstream gaming studios. In the present analysis of Tidvis, we also have unique access to the studio's 'inner life' since one of the authors of this article is the company's CEO and lead designer of *The Widow's Boutique.*

4. Methodology

In her work, Bódi has convincingly reconstructed the design ethos of three game developers. The three cases collectively demonstrate variations in how a design ethos may be understood and described. Each case study begins by setting the context of the studio's history, its evolution, and its key games, which provides a foundation for understanding the studio's design philosophy. Bódi discusses Naughty Dog's transition from platformers to cinematic experiences, highlighting how their design ethos evolved to prioritise narrative control and cinematic quality over player freedom. Similarly, BioWare's analysis traces their shift from traditional role-playing games to more action-oriented narratives, reflecting changes in design ethos influenced by market trends and internal team dynamics. System Era Softworks' chapter, on the other hand, explores their indie roots and how their design ethos fosters player creativity and exploration in sandbox environments. Through these three cases, Bódi demonstrates how *design ethos* serves as a framework for understanding how studios communicate their creative identity and design philosophy through promotional materials, industry coverage, and the aesthetic qualities of their games over time.

In formulating Tidvis' design ethos, we have strived to follow Bódi's method. However, where her work focused on player agency (and developed a heuristic framework to that end), we are interested in aesthetics and historical authenticity when seeking to unveil a history game design ethos. In accordance with Bódi, we begin by outlining the studio's history, including its founding milestones, some of its early work with games, and how its focus on aesthetics and historical accuracy has evolved. The basis for this work has been to analyse paratexts in the form of written and oral accounts of and about the company and some of its projects (Genette 1997; Consalvo 2017).

The paratextual analysis is based on information from the studios and the games' web pages, a report from a previous project, and three interviews with the CEO of Tidvis. The first is a 90-minute semi structured interview conducted in February 2018 about projects they were working on at the time (i.e., years before the ideas for *The Widows Boutique/Fru Sembs Valg* were conceived).[1] The second interview is a one-hour streamed conversation about games as a means of historical dissemination at the Bergen Public Library in November 2023 (Bergen Offentlige

---

1   Hutchison, R. (2018). "Interview by Magnus Henrik Sandberg." (19 February 2018)

Bibliotek 2023). The third is taken from an episode of the podcast Spillpedagogene (The Game Pedagogues) from October 2023 (Spillpedagogene 2023), in which Sandberg is one of the hosts.

Hutchison's central role in the studio and the games studied, combined with her participation in writing this article, implies that this research involves *insider research* positioning (Fleming 2018). While this provides valuable deep insights into the development process, we acknowledge the need to manage potential bias from this close involvement carefully. We recognise that our analysis does not simply uncover an existing design ethos but rather constructs an interpretation from complex data. To ensure research trustworthiness (Creswell 2013), we implemented a structured process in which Sandberg and Brazier wrote the initial analysis, which Hutchison then reviewed and elaborated upon before again leaving it to Sandberg and Brazier to finish the analysis. Following Patton's (1990) guidance for "logical, traceable and documented" (294) research, we have carefully grounded all interpretations in our paratextual analysis.

5. Analysis

In the following analysis, we will start by probing into Tidvis' design ethos. Then, we shall turn to *The Widow's Boutique* and analyse the game with regard to the studio's design ethos. Finally, we will discuss the findings of both previous parts in light of theory and previous research.

5.1 Tidvis' design ethos

A key to understanding the design ethos of Tidvis may be found in a project that preceded the game by more than ten years and which may also be regarded as the root of *The Widow's Boutique*. Before Tidvis was established as a company, the central founders brought 18th-century Norwegian trade to life by digitising historical customs and ship arrival records (Hutchison 2017). The team transformed over 56,000 handwritten entries from 1686-1794 into searchable data. They meticulously modernised and standardised place names and goods while preserving the wide variety of 18th-century descriptions. Later, they created other databases by digitising handwritten records in a similar way (Tidvis n.d.b).

Although it took some time before the group of independent historians that were later to form Tidvis created their first game, the project's playful ambitions shone through in diverse kinds of dissemination of this project's findings. In addition to the development of teaching material to be used in school and public appearances in seminars and various events targeted at the history-interested public, the customs records project spawned the five-episode radio series *Sensing the 1700s* on NRK (Norwegian Broadcasting Corporation), accompanied by a free, downloadable exhibition with the same name that has been displayed at more than 25 different venues spanning from schools to established museums. On TV, they contributed to NRK's history-themed reality show *Anno* by providing insights into 18th-century food and trade practices. The project's collaborative spirit extended to the public through a Local History Wiki-project run by the Norwegian National Library, where users contributed to developing knowledge about individual customs houses and trade patterns (Lokalhistoriewiki n.d.).

Although several of these project outcomes must be said to be quite playful compared to the more traditional ways of disseminating historical findings, throughout the project, an ideal of historical accuracy remained paramount. This combination of rigorous scholarship and public engagement earned the project the Researcher Union's Brain Power Prize in 2014, recognising its significance for both academic and public understanding of Norwegian history (Stolte vinnere av Hjernekraftprisen 2014, n.d.).

Tidvis' first game was also an outcome of the customs list project. *Privilegert*, released in 2016, was a physical *megagame* played on a board but with digital elements, designed to be played in a classroom with no less than 12 players.[2] Players took the role of either farmers, shipowners, or wealthy merchants in a game that simulated the region's actual export trade of that era. In the interview from February 2018, which happened in the context of a planned (though as of 2025, not yet accomplished) project of developing a fully digital version of the game, a number of issues relevant to sketching the design ethos of Tidvis were discussed. These entries, as well as the fact that the release of *Privilegert* marked Tidvis' debut as a game-developing studio, make it interesting to include the interview in the source material for the present analysis.

The stated reason for developing *Privilegert* was that Norwegian 18th-century economic history was often taught in uninspiring ways. *Privilegert* was introduced as a 'dissemination project', granting access to quantitative data from the customs list project about import, export and prices. These elements are put into use in the game and are, according to the developer, what makes this game stand out from other 'so-called historical games'. They elaborate by claiming that *Privilegert* is made by historians who have not been 'taking shortcuts to get to the drama'. While representing a rather harsh and general judgment on other historical game developers, these statements also give a clear indication of Tidvis' self-perception as a disseminator of historical research as much as a game developer.

As this happened in the early years of Tidvis, the company contracted an interactive game design company to help create the mechanics for this first game. Still, the studio's dedication to history triumphed over gameplay considerations when there was a discrepancy between the two, and the game is thus closer to a simulation than one might imagine it could be. Prices on goods in the game are genuine,

---

2   See *Privilegert*'s website. Retrieved from https://www.privilegert-spill.no/.

as are the price fluctuations, which are also central to the dynamics that make decisions difficult and interesting later in the game.

This dedication to the historically correct went so far as to set the game's balance in peril when disagreements between Tidvis and the external game development company were settled by historians of Tidvis reasoning that "we know this happened". The ethos of staying true to history is also reflected in the *About* section of the game's website, where it is said that during piloting sessions:

"[W]e have seen several cases where people have developed highly sophisticated financial instruments in attempts to secure the future of their farm, trading house, or shipping company. We have let players do so, as this also happened in the 1700s. Then, as now, it has sometimes worked out well, sometimes not." (Om spillet n.d.)

Somehow, this reasoning could also be seen as a foreshadowing of the digital game *The Widow's Boutique*, where smuggling is a key component. What may be regarded as breaking the rules can just as well be seen as genuine simulations of economic mechanisms. It follows from the commitment to authenticity that the rules of the board game cannot be clean and clear if the rules of the trade that the game depicts are not.

Privilegert's emphasis on authenticity is further shown by the production of three distinct versions of the game, each tailored to the trade practices of different regions of Norway and illustrated using 18th-century genre art from the different regions, most of which have not commonly been depicted in popular culture. In the Eastern Norway and Southern Norway versions of the game, the trade is in lumber, and in the Western Norway version, they trade in fish. A fourth version was planned to focus on the copper export of Mid-Norway, but it was never finalised. The stated reason for cancelling it was that Tidvis found the economic structure of the early modern copper trade to greatly favour one group of actors, making trade negotiations impossible. Again, what *could have been* a decent strategy game was aborted as balancing it into a fair and fun game to play would have jeopardised what the designers see as historically accurate. The decision to cancel the copper trade version reflects Tidvis's commitment to historical accuracy over gameplay convenience. Rather than abandon this story entirely, they later developed a museum exhibition about copper export, demonstrating their dedication to finding appropriate formats for different historical narratives (Tidvis n.d.d).

At the time of the interview, in 2018, Tidvis had an ongoing project called *Oslo havn 1798*, where they attempted to digitally recreate the harbour area of Oslo. Tidvis thought of the project as a test of what could be accomplished when historians, game developers, and illustrators worked together Illustrators drew the houses, and historians transcribed insurance documents that contained descriptions of each house along the street. Sometimes they had drawings, and sometimes they had to rely on their expertise on the building technique used. At the time of the interview, however, Tidvis did not know how the digital model was to be used. They just made it, then made it publicly available, hoping that it would be useful to someone. "If they decide to make an *Assassin's Creed 1798* set in Oslo, we will be the experts they need to consult"[3], they said jokingly.

The strength of games, the CEO of Tidvis says, is that you "feel it in your bones"[4]. Still, talking about *Privilegert*, she holds that the game may easily lose its value if the players are not properly debriefed afterward. On a surface level, this reflection may lead to *Privilegert* being understood as a game meant for use in a school context for knowledge transfer (Squire 2002). However, the essential point is rather that Tidvis sees history games as mediating historical insights that are not essentially connected to this (or any) particular medium. If, after playing, the theme is talked about and contextualised differently, one may be able to peel off the media-specific attributes to more clearly see what Tidvis regards as the essence of Norwegian 18th-century economic history. Contrarily, to play *Privilegert* and not talk about it afterward would lead to lost opportunities to reflect and learn about the past.

This emphasis on post-play reflection aligns with Jörn Rüsen's conception of historical consciousness as a process that involves more than knowing historical facts; it entails ethical orientation and critical engagement with the temporal dimension of human experience. For Rüsen, historical learning becomes meaningful when past events are interpreted in ways that help us make sense of our present and guide our actions. From this standpoint, the act of playing the game alone is not sufficient – it is in the contextualisation and discussion after play that the mimetic experience can be transformed into historical consciousness.

Tidvis' design ethos is defined by a steadfast commitment to historical authenticity, rigorous scholarship, and public engagement. Emerging from collaborative archival projects, the studio's approach is rooted in a historian's sensibility: primary sources are not simply referenced but drive both content and mechanics. This ethos is evident in their willingness to prioritise historical accuracy, sometimes at the expense of game balance or entertainment value, and in their belief that games can be vehicles for disseminating research, not just storytelling or play. By treating the game as a medium for historical dissemination, Tidvis positions itself less as a traditional game developer and more as a facilitator of historical consciousness, where critical reflection and ethical engagement are as integral to the experience as the gameplay itself.

---

3   Hutchison, Interview.
4   Hutchison, Interview.

5.2 The Widow's Boutique – historic authenticity through visuals and mechanics

We will now see how the design ethos of Tidvis is reflected in one of their most recent games. *The Widow's Boutique* is Tidvis' first game in English. The game is a translated and slightly altered version of a previously released game in Norwegian called *Fru Sems valg* (*Mrs. Sem's Choice*). As they are practically the same game with a shared design history, we will draw on both games and see their paratexts as shared in this analysis. We will point to the differences between them where applicable. The original version in Norwegian received three prizes at the Norwegian annual games award Spillprisen 2022. It is free, published on an open license CC-By-SA, and available on Google Play and App Store, as well as playable in a browser on a PC.[5] The global version, available in English, is sold commercially and available on Google Play, App Store, and Steam.

In *The Widow's Boutique*, players embody the protagonist, a recent widow tasked with managing her deceased husband's shop. The game is set in the winter of 1821, and in addition to running the shop, the player is, from the very beginning, introduced to people and gets to participate in the social and cultural life of Christiania. The role of Mrs. Sem emphasises personal agency and resilience in a male-dominated society, confronting challenges concerning business, romance (which is also business), and social life (which is also business).

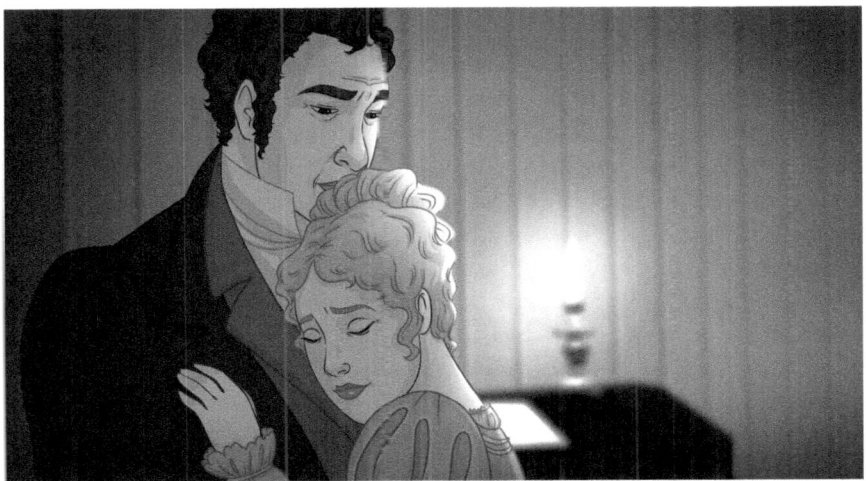

Fig. 1: *The possibility of a romance with Customs Officer Bush shows how The Widow's Boutique blends the romantic theme of visual novels with the business strategy theme.*

---

5 See *Fru Sems valg*'s website. Retrieved from https://frusemsvalg.no.

Tidvis describes *The Widow's Boutique* as a "Nordic noir historically accurate Visual novel" (The Widow's Boutique n.d.). As a genre, the visual novel represents a distinctive form of interactive digital storytelling that merges narrative text with static or animated illustrations, typically featuring anime-inspired artwork (Cavallaro 2009). They emphasise rich storytelling over complex gameplay mechanics, with the primary interaction consisting of advancing through text while viewing character portraits and scenic backgrounds. Players shape the narrative through choices that branch the story into multiple possible paths and endings. These branching narratives are among the features that make visual novels potentially interesting as tools for learning (Øygardslia et al. 2021). The genre encompasses a broad spectrum of themes and styles, from romance and drama to science fiction and horror. While most titles originate from Japan and are associated with anime, Tidvis' use of the term *Nordic noir* places *The Widow's Boutique* in a distinct Nordic tradition of suspense storytelling and associates it with other forms of narrative media such as literature and film.

In adherence with the visual novel genre, *The Widow's Boutique* displays 2D illustrations that usually remain static throughout a scene (Cavallaro 2009). While the genre conventions of visual novels can be said to provide relief from expectations of photorealism, the designers of *The Widow's Boutique* are committed to faithfully reconstructing historical visual environments. Clothes, hairstyles, furniture, and tapestry all accord with the fashions of the time and are based on research into museums and other collections, in cooperation with art and fashion historians. Likewise, the game narrative authentically portrays early 19th-century social norms and etiquette. A key example is how marriage affects the main character's fate: if the widowed protagonist chooses to remarry, she must surrender control of her business to her new husband, and the game ends. Thus, it forces the player to experience historical gender norms first-hand, as the narrative-driven consequences of marriage reflect real social structures, allowing the player to both engage with history and develop a historical consciousness rather than just be presented with historical facts (Rüsen 2004).

Indeed, this historically accurate consequence resonates with the real-life experience of the woman who inspired the main character. While Mrs. Else Marie Strøm did not remarry, she was forced to hand over the store (which is today Oslo's oldest department store, Steen & Strøm) to her son when he grew old enough. During her approximately ten years of running the business, she was the one who introduced the textiles, for which the store later became famous. However, even if Mrs. Sem's boutique also specialises in textiles in the same city and the same time period, it is important for Tidvis to emphasise that the game does *not portray* Mrs. Strøm.

Contrary to *Privilegert*, where the players take one of three different roles of economic actors in 18th century Norway and thereby clearly take part in a modelling of an economic system, *The Widow's Boutique* puts the player in control of an individual named character who interacts with others and makes decisions which can sometimes be ethically questionable. To Tidvis, this represented a problem, and it was detrimental to the company that these characters were not understood as portraits of real contemporary persons. They chose to clarify this on the webpage of *Fru Sems valg*: "The dramatic parts of the games are fictional but inspired by history. This applies to the protagonist and most of the main characters you meet. The reason is that we do not want to attribute attitudes or actions to people that we are not certain they had or did." (Om oss n.d.)

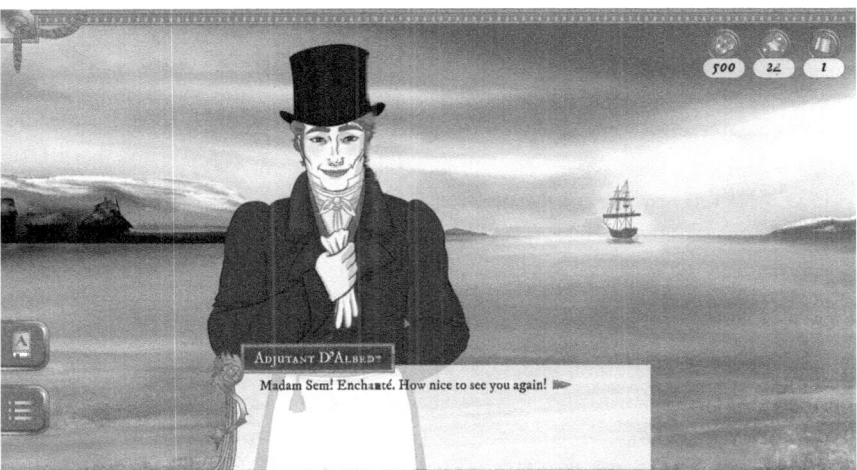

Fig. 2: *The visual style of the game affords opportunities to show accurate clothing design.*

Despite this reluctance to portray real people, there are exceptions. Playing as the fictional Mrs Sem, the player may visit her dear friend Conradine Dunker, who runs a girls' school in her own home. Mrs. Dunker was a well-known writer of the day whose memoirs have been re-published several times. That makes it easier to study her writing and author dialogues that are in line with the attitudes she expressed. The archives have very limited sources on Mrs. Strøm. In addition, it also seems a lot easier to include a historical person in a supporting role than as the playable protagonist. The player gets to speak to Mrs. Dunker on several occasions but never *as* her. Mrs. Dunker's function in the game is to provide information as she responds to things the player, in the role of Mrs. Sem, chooses to tell or ask her.

When speaking with Mrs. Dunker in the Norwegian version of the game, the dialogue box has an icon linking to the Norwegian Biographical Encyclopedia's article about Conradine Dunker. This way, the player is encouraged to learn more about this social entrepreneur who lived in Christiania at the time. There are several of these external links in the game, for instance, from the daily newspaper articles, which are genuine, to the original paper of that day available digitally in the National Library. In addition to facilitating further exploration of people or phenomena being linked to, these links build a form of authenticity by documenting that these particular game elements are accurate.

These links are not included in *The Widow's Boutique*, partly due to the sources being in Norwegian and only available to Norwegian IP addresses, and partly because of the interruption to the game flow they may trigger. This difference between the game versions may, however, also indicate a difference between a game development funded through grants and a game offered to the commercial consumer market. The former may have a greater interest in displaying its historical relevance and its usefulness in a history-learning context.

A different source of historical information, which can be found in both versions of the game, is the internal encyclopedia of fabrics. Upon visiting the store, the player, as Mrs. Sem, may take care of customers by recommending suitable fabric for their needs. The game mechanic is a simple quiz, but the answer can be looked up in the encyclopedia. Giving the customer the right recommendation has repercussions on the character's status, as well as future customers and income. Thus, engaging with customers gives the player both the pleasure of interacting with the société of Christiania and learning more about the trade. In addition, picking the right fabric adds both to the store's immediate income and its reputation. Thus, it both speaks to the pleasure of playing the role well and the strategic means of winning the game. Naturally, the 15 different fabrics one may learn about and choose from were all imported to Norway at the time. Thus, what first appeared as a line of handwritten text in the original customs records is given a visual form and function in the descriptions of the encyclopedia and embedded in the game story as the needs of an inhabitant of Christiania as well as in the business mechanics.

While the audiovisual style, as well as the social conventions expressed through verbal language, offer a realistic depiction of early 19th-century Christiania, the game also makes room for a different mechanic that is less typical of visual novels and which offers a different take on historical understanding and authenticity. Namely, the economic system. Buying and selling premium manufactured textiles could be both difficult and risky in addition to economically, socially, and ethically challenging. It also involved a number of procedures which, when re-enacted by a player, can offer insights and be both historically authentic and playable. Thus, authenticity in historical games can be found in the procedures (Bogost 2007). The mechanics of running the store itself are more typical of a business simulation game than a visual novel.

Connected to the business simulation, *The Widow's Boutique* also examines how marriage was fundamentally an economic arrangement, where women's choices were limited by financial necessity and social pressure. As previously mentioned, if the player, as Mrs. Sem, decides to marry one of her suitors in the game, the suitor takes over the business, and the game ends. More general examples of the game's socially realistic storytelling are the debtor's prison, which the player ends up in if unable to pay her debt, the poor children that pickpocket the player, and the servant girl whom you can discover stealing from the till. The game is thus a combination of social realism, possibilities for romance, and a strategic business simulation, all complemented by authentic fashion styles and interior design.

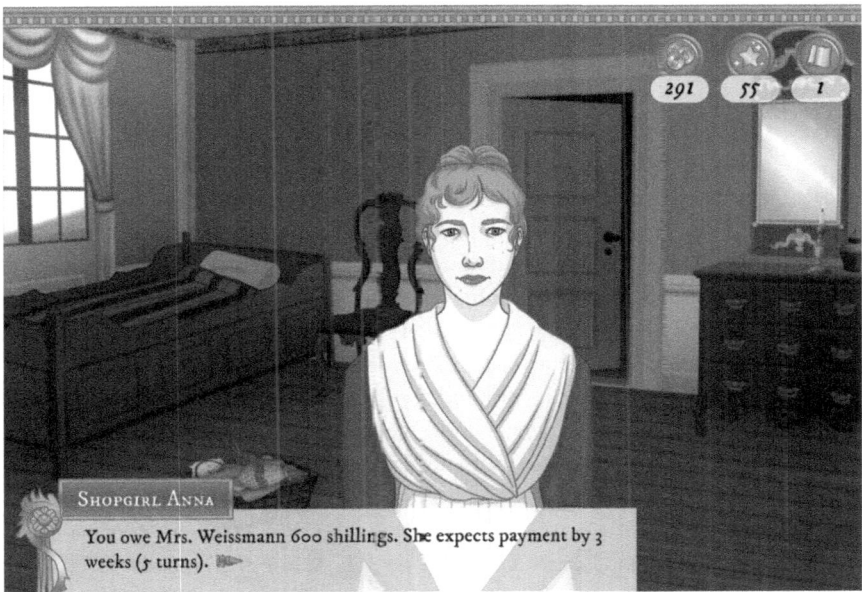

Fig. 3: *The business simulator mechanics are interwoven in the narrative. The shop girl reminds you of the economic challenge, and the key numbers are always visible in the top right corner.*

This way, *The Widow's Boutique* exemplifies Tidvis' philosophy by weaving meticulous historical research into both its visual presentation and core mechanics. The game's commitment to authenticity is visible in details such as period-accurate clothing and interiors, as well as the economic and social constraints embedded in gameplay. By situating players within the lived realities of 1820s Christiania, the game enables a nuanced engagement with history: players experience the consequences of gendered laws, economic hardship, and social norms firsthand. The integration of branching narratives, business simulation, and educational resources transforms the game into both a compelling narrative and a critical historical document, demonstrating how interactive media can foster historical understanding and ethical reflection beyond mere entertainment.

6. Discussion

This analysis shows how *The Widow's Boutique* is both a product of the design ethos of the studio that made it and, in turn, helps to evolve it. A key focus is on the tension between historical fidelity and the imperatives of playability.

In line with Copplestone's findings about 'cultural-heritage professionals', the design ethos of Tidvis displays an ambiguous relationship with the question of accuracy. While their own dissemination of tools for historical engagement over several years and many projects is of a playfully constructionist or even deconstructionist kind, they tend toward reconstructionism and focus on accuracy and what 'really happened' when talking about it. In commenting on history games in general, they seem to be even more strict, insinuating that other developers are "taking shortcuts to get to the drama".[6] Whereas this critique is vague and perhaps unfair, the expressed ambiguity may be understood as reflecting the dynamic interplay between facts and interpretation typical of historical realism (Carr 1961). It could also be read as an attempt to safeguard professional perspectives in relation to the selective authenticity sometimes associated with developer-historians (Southgate 2005; Salvati/Bullinger 2013; Chapman 2018).

That said, Tidvis' attention to historical accuracy does not mean that their games cannot contain fictional elements. In the making of *The Widow's Boutique*, the decision to create a fictitious main character was taken after considering making the game about a real historical person. While the story is based upon and, in some ways, closely knit into the biography of a real historical character, the team concluded that sources about her were too scarce and that making undocumented claims about an individual was unethical. Tidvis' early collaboration with an external partner for *Privilegert* is revealing. When the partner proposed to balance tweaks that would have smoothed out the historical price shocks, Tidvis' historians over-ruled them, arguing that "we know this happened". This micro-struggle exemplifies Adorno's notion of resisting the culture industry's demand for frictionless fun and, in Bódi's terms, signals a design ethos where scholarly authority can veto market-conventional advice.

Tidvis' decision to use a fictional protagonist in the game aligns with Rüsen's concept of historical consciousness, which emphasises how history should be ethically responsible, critically engaged, and structurally authentic (Rüsen 2004). Rüsen argues that historical consciousness develops when individuals actively engage with history rather than passively consume it. Tidvis has retained authenticity through an accurate representation of social and economic constraints, which allows players to experience historically grounded dilemmas without misrepresenting real individuals like Mrs. Strøm, but at the same time include

---

6 Hutchison, Interview.

real historical actors such as Mrs. Dunker in order to include multiple perspectives.

This indicates another interesting and more general point when creating games for history dissemination. Painting a historical portrait of a real and identifiable person in the form of a game is, in certain ways, more challenging than portraying economic or other structural matters in this medium. In the latter, as exemplified in *Privilegert*, the point for the player may be exactly to explore the possibilities within a given framework. While there is, of course, room for interpretation also at the individual level, a game that retells a historical person's life story is more constrained by what actually happened. That may be an explanation as to why real historical characters tend to be sidekicks in games and not the main characters controlled by the player. Equally telling is Tidvis' decision to leave ethically dubious options – smuggling silk, bribing customs officials, cutting servants' wages – fully playable. Rather than punish or preach, the game lets players succeed or fail on historically plausible terms. That design choice produces what Adorno calls dialectical mimesis: complicity is felt, not sermonised. For Rüsen, such ambiguity is a prerequisite for ethical judgement because it forces players to interrogate how far economic pressure excuses unethical acts.

A consequence of Tidvis' clearly stated and demonstrated ambition of historical accuracy is that their games may also be criticised as such. This stands in contrast to historically themed games which have as their primary aim to entertain; one may always explain inaccuracy by referring to this aim. This criticism can to some extent be leveled at *The Widow's Boutique* and highlights the difference between the Norwegian and English versions of the game. The removal of links to external resources that added to the historical authenticity and depth in *Fru Sems valg* seems to have been traded for accessibility in the commercial adaptation. Although this move was a pragmatic adjustment to the technical and language challenges connected to keeping these in-game links, the developers also argued that they were removed to prevent game-flow disruption. This reflects, at least partially, a shift from critical historical engagement to player experience. The omission reflects Consalvo's (2017) argument that commercial game design often prioritises engagement over historical accuracy but also echoes Adorno's warning that authenticity can be sacrificed for entertainment or historical spectacle (Adorno/Horkheimer 1997 [1947]). The link removal also exposes the commercial fault line in Tidvis' ethos. Hyperlinks are paratexts that foreground evidence and invite verification, but they also disrupt the seamless flow prized on Steam and the App Store. Their excision, therefore, marks a point where the culture-industry logic Adorno critiques reasserts itself, reminding us that even avowedly scholarly studios must negotiate the market's demand for frictionless consumption.

Much of the developer's claim regarding the game's accuracy, which is central to its authenticity, rests on its relation to the primary sources it is built upon, particularly the databases of customs and ship arrival records from the period. However, these sources are not directly accessible to players through an in-game interface and thus do not distance players from the historical context as other database structures in historical games have been claimed to do (Schoppmeier 2023). Instead, the databases' significance and authority are communicated through non-diegetic channels such as the game's website, interviews, and related dissemination projects, which many, but likely not all, players will encounter. To truly recognise *The Widow's Boutique* as a historical game, one must accept that the history it presents is as significant as the gameplay itself. This raises the question of whether *The Widow's Boutique, Fru Sems valg*, and *Privilegert* should all be seen as paratexts of the databases they disseminate (Consalvo 2017).

When the basis for the historical narrative lies in rather obscure primary sources rather than well-known events, a different process and a different tone in the narrative is in place. From the viewpoint of game studies, Tidvis takes the consequence of these challenges by not necessarily creating games at all. *The Widow's Boutique* is but one of a number of creative dissemination projects that build on the same historical databases. Several of the other projects are not even games. There are 3D simulations, museum exhibitions, smell boxes, and a cookbook (Tidvis, n.d.c). In the specific context of Christiania in the 1820s, there are aspects of the depiction of society which does not easily fit in with the narrative of *The Widow's Boutique*. In the game, we see hints of poverty, alcoholism, and even concealed birth and infanticide. However, while these stories must be told, they fit badly with the tone of this game's narrative. Tidvis then chose to disseminate these parts of history in other games or mediated them altogether differently. In other words, this is not a case of a game development company deciding on the theme of their next game, but of historians considering how to present their data in the most engaging way.

7. Conclusion

Through the combination of analysing the game's aesthetics and reconstructing the design ethos of Tidvis, we hope to have shown how *The Widow's Boutique*, continues, refines and expands upon the studio's long-standing engagement with primary historical sources and critical dissemination practices. In doing so, the game keeps in clear view the very historical questions underlined in the introduction – namely, how women negotiated commercial agency and how 19th-century Christiania embodied the wider currents of emergent capitalism in Europe. As this study has demonstrated, the historical data used in the game is not confined to *The Widow's Boutique* alone but is part of a broader ecosystem of historical dissemination – used in earlier projects like *Privilegert*, as well as in museum exhibitions, databases, and even non-game formats like cookbooks. In this sense, the game must be understood not in isolation but as part of a wider network of historical storytelling, each version and medium offering a different lens through which to understand the past.

This expansive and layered practice of history underscores a crucial point: Tidvis' games are not self-contained experiences. Questions of historical accuracy and authenticity quickly give way to discussions about the relevance of historical perspectives beyond the game itself. In Tidvis' framework, a gaming session is rarely an end point – it is instead an invitation to explore further. The real value of the game unfolds in the reflections, conversations, and contextualisation that follow. In this way, their games function as what Consalvo (2017) terms paratexts – but also as prompts for ethical engagement, in line with Rüsen's model of historical consciousness, which holds that historical narratives must guide moral reflection and multi-perspectival understanding.

Seen in this light, *The Widow's Boutique* does more than reflect Tidvis' ethos – it advances it. The game builds upon earlier projects such as *Privilegert*, not only in terms of content but through a fusion of genres. By combining the aesthetic and emotional intimacy of the visual novel with the economic procedurality of business simulation, the game models both the constraints and the lived experience of early 19th-century life. This hybrid design is not simply a formal experiment; it enables a type of *critical authenticity* – a term that moves beyond visual fidelity to capture how the game encourages the player to feel and reason within a historically situated system of social, economic, and ethical tensions.

Crucially, *The Widow's Boutique* leverages the visual novel format to create an affective environment that avoids the reduction of history to tropes or spectacles. While it is a game grounded in visual presentation, it resists the commodified aesthetics that Adorno warns against, in which mimesis is flattened into market-friendly replication. Yet this resistance is not total. As the link-removal episode shows, Tidvis must still negotiate the usability norms of digital storefronts. The concession does not nullify their critical intent, but it reminds us – as Adorno would – that every historical game remains entangled in commodity circulation. Instead, the game crafts a nuanced form of mimesis – one that engages players in dialectical reflection rather than passive consumption. It asks players not simply to observe history but to inhabit it: to feel the pressure of social norms, to weigh uncertain choices, and to experience the consequences of decisions that are historically plausible, ethically fraught, and narratively rich.

All of this affirms that Tidvis is not only striving for historical accuracy but is treating history as a living, ethical, and procedural domain. Indeed, their strength may be that they do not primarily regard themselves as game developers at all but as historians with a passion for finding the most meaningful ways to disseminate the past. With their feet firmly planted in archival sources and their eyes set on public engagement, Tidvis has carved out a space at the intersection of historical scholarship, game design, and cultural pedagogy. More broadly, the case suggests that a *historical game design ethos* is defined by three intertwined commitments: the transparent use of primary sources, the crafting of mechanics that stage rather than obscure structural constraints, and the deliberate creation of post-play pathways for ethical and scholarly reflection.

References

Adorno, T. W. (2014 [1970]): Aesthetic Theory. Paperback ed., Repr. Bloomsbury Revelations. London: Bloomsbury.

Adorno, Theodor W./Max Horkheimer. (1997 [1947]): Dialectic of Enlightenment. London: Blackwell Verso.

Apperley, T. (2010): Gaming Rhythms: Play and Counterplay from the Situated to the Global. Amsterdam: Institute of Network Cultures.

Bergen Offentlige Bibliotek. (2023): "Fru Sems Valg - Dataspill Som Historieformidler". [Video file]. Retrieved from https://www.youtube.com/live/aqCqyFpJ4DA?si=Gnu2qQMFiKjuVa3J.

Bogost, I. (2007): Persuasive Games: The Expressive Power of Videogames. Cambridge: MIT Press.

Bódi, B. (2022): Videogames and Agency. London: Routledge.

Carr, E. H. (1961): What Is History?: The George Macaulay Trevelyan Lectures Delivered in the University of Cambridge, January-March 1961. Princeton: Macmillan.

Cavallaro, D. (2010): Anime and the Visual Novel: Narrative Structure, Design and Play at the Crossroads of Animation and Computer Games. Jefferson: McFarland & Co.

Chapman, A./Foka, A./Westin, J. (2017): "Introduction: What Is Historical Game Studies?" Rethinking History 21(3), pp. 358–371.

Clyde, J./Hopkins, H./Wilkinson, G. (2012): "Beyond the 'Historical' Simulation: Using Theories of History to Inform Scholarly Game Design." Loading... The Journal of the Canadian Game Studies Association 6(9), pp. 3–16.

Consalvo, M. (2017): "When Paratexts Become Texts: De-Centering the Game-as-Text." Critical Studies in Media Communication 34(2), pp. 177–183.

Copplestone, T. J. (2017): "But That's Not Accurate: The Differing Perceptions of Accuracy in Cultural-Heritage Videogames between Creators, Consumers and Critics". In: Rethinking History 21(3), pp. 415–438.

Cresswell, J. W. (2013): Qualitative Inquiry and Research Design: Choosing among Five Traditions. Thousand Oaks: Sage.

Fleming, J. (2018): "Recognizing and Resolving the Challenges of Being an Insider Researcher in Work-Integrated Learning." International Journal of Work-Integrated Learning 19(3), pp. 311–320.

Genette, G. (1997): Palimpsests: Literature in the Second Degree. Lincoln: University of Nebraska Press.

Hammar, E. L. (2023): "Producing and Exploiting the Cultural Memory of the American West. Imperialism, Gender, and Labor at Rockstar". In: J. Wills/E. Wright (eds.), Red Dead Redemption: History, Myth, and Violence in the Video Game West. Norman: University of Oklahoma Press, pp. 149–163.

Hutchison, R. (2017): "Historiske toll- og skipsanløpslister– tollregnskap som kilde." Heimen 54(3), pp. 275–287.

Kapell, M./Elliott, A. B. R. (eds.) (2013): Playing with the Past: Digital Games and the Simulation of History. New York: Bloomsbury Academic.

Lebowitz, J./Klug, C. (2011): Interactive Storytelling for Video Games: A Player-Centered Approach to Creating Memorable Characters and Stories. Burlington: Focal Press.

"Lokalhistoriewiki" (n.d). Retrieved from https://lokalhistoriewiki.no/wiki/Lokalhistoriewiki:Hovedside.

Øygardslia, K./Weitze, C. L./Shin, Y. (2021): "The Educational Potential of Visual Novel Games: Principles for Design." Replaying Japan 2, pp. 1–12.

Om oss (n.d.). Retrieved from https://frusemsvalg.no/om-oss/.

Om spillet (n.d.). Retrieved from https://www.privilegert-spill.no/om-spillet-2/.

Patton, M. Q. (1990): Qualitative Evaluation and Research Methods. Newbury Park: Sage.

Rahimi, F. B./Kim, B./Levy, R. M./Boyd, J. E. (2020): "A Game Design Plot: Exploring the Educational Potential of History-Based Video Games." In: IEEE Transactions on Games 12(3), pp. 312–322.

Rüsen, J. (2004): "Historical Consciousness: Narrative Structure, Moral Function, and Ontogenetic Development." In: P. C. Seixas (ed.), Theorizing Historical Consciousness. Toronto: University of Toronto Press, pp. 63–85.

Salvati, A. J./Bullinger, J. M. (2013): "Selective Authenticity and the Playable Past." In: M. Kapell/A. B. R. Elliott (eds.), Playing with the Past: Digital Games and the Simulation of History. New York: Bloomsbury, pp. 153–67.

Schoppmeier, S. (2023): "The Gameworld, the Interface, and the Genre." In: J. Wills/E. Wright (eds.), Red Dead Redemption: History, Myth, and Violence in the Video Game West. Norman: University of Oklahoma Press, pp. 27–44.

Southgate, B. C. (2005): What Is History For? London: Routledge.

Spillpedagogene (n.d.): "Spillpedagogene: En Podcast Om Spill Og Digital Kultur". Retrieved from https://open.spotify.com/episode/1J3C5bwkBOPSCEWLcION5Y?si=4IhWYaW7STq78-GbDDEKrQ.

Squire, K. (2002): "Cultural Framing of Computer/Video Games." Game Studies 2(1), n. p..

"Statement on Good Practice for Historians" (n.d.). Retrieved from https://royalhistsoc.org/policy/position-papers/statement-on-good-practice/.

"Statement on Standards of Professional Conduct" (7 January 2023). Retrieved from https://www.historians.org/resource/statement-on-standards-of-professional-conduct/.

"Stolte vinnere av Hjernekraftprisen 2014" (30 January 2015) Retrieved from https://www.forskerforbundet.no/om-forskerforbundet/hjernekraftprisen/vinnere/stolte-vinnere-av-hjernekraftprisen-2014.

The Widow's Boutique (n.d.). Retrieved from https://widowsboutique.com/.

Tidvis (n.d.a): "Hvorfor". Retrieved from https://tidvis.no/#hvorfor.

Tidvis (n.d.b): "Historiske Databaser". Retrieved from https://tidvis.no/historiske_databaser/.

Tidvis (n.d.c): "Hvordan". Retrieved from https://tidvis.no/#hvordan.
Tidvis (n.d.d): "Museum". Retrieved from https://tidvis.no/prosjekter/museum/.
Weiss, J. (2011): The Idea of Mimesis: Semblance, Play, and Critique in the Works of Walter Benjamin and Theodor W. Adorno. Chicago: DePaul University.

Enchanted Imaginings
Involving Museum Visitors in Heritage Adventures

Eva Kingsepp and Linda Ryan Bengtsson

Abstract

In a heritage context, physical objects as well as digitally produced representations can fulfill important functions in people's experienced relationship to the past, but we need to investigate the actual foundations for such experiences. How can we understand the combined roles of objects, narrative, and spatial location in visitors' feelings of connectedness to place-specific historical events? The article presents a model of what happens at the intersection of place, object(s), technology, visitor(s), and historical narratives, proposing imagination and play as a valuable, often neglected resource. The model builds on augmented reality/AR concepts developed in collaboration with visitors, artists, stakeholders, museum personnel and researchers for a heritage site, Långban, Sweden. Here, physical objects and digital technology were combined to involve visitors as co-creators of emotionally and intellectually captivating narrative game-based experiences. In such a setting fact and fiction meet. The imaginative is prompted by digital media interconnecting the physical (architecture, landscape, and objects), spatial movements, game mechanisms and narratives. We argue that it is potentially rewarding to make use of the visitor's own imagination in the co-creation of an enhanced/enchanted experience at heritage sites or museums, as this can allow for a high level of imaginative engagement that personally embeds visitors in the narrative, beyond factual communication.

Keywords

digital heritage, open-air museum, gaming, augmented reality, immersion

1. Introduction

"Is it not possible [...] that things we have felt with great intensity [...] have an existence independent of our minds; are in fact still in existence? And if so, will it not be possible, in time, that some device will be invented by which we can tap them? [...] Instead of remembering

here a scene and there a sound, I shall fit a plug into the wall; and listen in to the past."
(Lowenthal 2015 [1985]: 55)

In *The Past is a Foreign Country* (2015 [1985]) historian and geographer David Lowenthal examined our relations to the past. Using travel as a metaphor in this context inevitably leads to the concept of *time travel* and fantastic imaginations about looking into the past – or even entering it – through a magic space-time wormhole. Today the potential of such experiences has been granted to immersive digital technologies. We find the metaphor particularly intriguing regarding museums located at the actual site of historical events, as the spatial connection is crucial in visitor experiences of authenticity and immediacy, promoting a feeling of what can be called enchanted immersion in digital place-based narratives about the past. In this article, we will focus on what is often referred to as augmented reality (AR).

AR is at its core an overlay of digital material onto the physical. The digital layer uses video, sound graphics, or data that contributes with additional information or narratives to physical space, enhancing (augmenting) the viewer's or user's experience (Laine/Suk 2016). Museums and heritage sites are increasingly exploring and developing digital productions such as AR to evoke audience engagement (Parry 2010; Bujari et al. 2016; Chiu et al. 2017; Javornik et al. 2019; Marchant/Nancarrow 2019). However, there are often shortcomings related to the design of digital solutions, not the least when impressive digital simulations are foregrounded at the expense of visitor engagement (e.g. Cameron/Kenderdine 2010; Braunerhielm/Ryan Bengtsson 2023). Moreover, as put by Bernadette Flynn, "interaction has often been used to mean a form of interpassivity where the user has limited options within a preprogramed menu. [...] Any approach that does not take user experiences as central will produce less than engaging spaces." (Flynn 2010: 354)

In the first part of this article, we present a theoretical framework for approaching visitors' engagement at heritage sites as 'time travel': as actually being there, transcending time at a particular place. There is an increasing body of academic literature addressing strategies for enhancing visitor engagement in museums, especially within participatory design that foster the co-creation of meaningful experiences with visitors (cf. Pruulmann-Vengerfeldt/Runnel 2018; Pierroux et al. 2020). This article, however, seeks to contribute with theoretical concepts useful when analysing, discussing and possibly developing the interplay between place, object(s), technology, visitors(s) and historical narratives at heritage sites. We do so by combining ideas from media studies, philosophy, literary theory, and game studies, and in particular the concepts enchantment and immersion, investigating how these apply to how engagement is formed at a site. We thus argue for the need to theoretically conceptualise how augmented technologies can intertwine place, narratives and imagination to enhance visitor experiences.

In the second part of the article, we apply the theoretical framework on a case study of AR concepts developed through participatory design with visitors, artists, stakeholders, museum personnel and researchers for an open-air heritage site at Långban, Sweden. The intention was to develop open-air museum installations grounded in the site's history and landscape (the design process is further developed in Ryan Bengtsson et al. 2022). Here local history, tangible objects and digital technology were combined to involve visitors and their imaginative capacities as co-creators of engaging narrative experiences.

Finally, we discuss how the concepts in our philosophical chain of thoughts work together when applied to an open-air museum. Our argument is that when place, digital tools (game and gamification) and narrative work together, they facilitate playfulness, thus potentially leading to a higher degree of visitor engagement.

2. History coming alive: Digital technology, place, and narrative

A decade ago, Refsland, Tuters and Cooley predicted that "location-based wireless devices will hypothetically permit an immersive experience in which users will be able to browse layers of digital information encoded to a particular place" (2010: 410). Since then, the relations of visitors to the combination of place, artefacts and digital technology have been discussed to some extent (cf. Lowenthal 2015; Ryan Bengtsson 2012; Champion/Dave 2010). However, as digital technologies are today becoming interwoven with places and actions, and – as some claim – even interconnected with how we understand and interact with a place (Couldry/McCarthy 2004; de Souza e Silva 2006; McQuire 2016; Fast et al. 2018; Hjort/de Souza e Silva 2023), this "mixing of code, data and physical place" (Thielmann 2010: 5) calls for new theoretical approaches.

When developing our conceptual setup we have in particular been inspired by research on narrative aspects in digital games, a media type frequently used to create engaging scenarios of virtual history. Such games often present speculative history narratives based on what-if, using actual places and characters as models for an entertaining, but also educating, simulation of past events (Uricchio 2011). In these games not only elements related to history are used, but also widely shared references to, for example, popular films, novels and comics, resulting in virtual worlds inhabited by both fictional characters and gamers' avatars (Kingsepp 2006; Kingsepp 2018). The same approach is found in Flynn (2010), with whom we also share our interest in experiences of immersion, authenticity, and enchantment, related to virtual representations of the past. Unfortunately, her otherwise well-informed and inspiring piece refers to these concepts without any definitions or other substantiated support. This is a common problem, and highlights the need of a distinct theoretical vocabulary in order to achieve a deeper understanding of the intersection of media and place.

Heritage as history coming alive, including aspects of performativity and entertainment, offers a rich well of resources for experiences of time travel, e.g. as seen in living history re-enactment in open-air museums (cf. Jordanova 2006; Samuel 2012; Lowenthal 2015 [1985]; Marchant/Nancarrow 2019; Sturani 2022). However, there is a significant difference between staged events within a pronounced institutional framework, where the visitor can be likened to a tourist in an organized simulation of the past (cf. Champion/Dave 2010: 333–334; Flynn 2010: 362), and a personal experience of immersive historicity (Kingsepp 2006): an overwhelming feeling of oneself actually being, and acting, in a hyperreal version of the past. As play and playfulness are widely acknowledged as fundamental for every human being, regardless of age (cf. Caillois 1958; Huizinga 1938), it is valuable to make more of this disposition, in particular as the heritage site itself can contribute significantly through providing a narrative framework and engaging local stories. Notably, allowing the place to take active part in the construction of the narrative, and the visitors' experiences of it, also means acknowledging the people for whom the place mattered as part of their everyday lives (cf. Lefebvre 1991; Braunerhielm/Ryan Bengtsson 2024). Cultivating a sensibility to place-based historical narratives, and through them an attachment to the people from the past with whom we share the site (cf. Heidegger 1987 [1935]: 37–38), is a way to handle the potential ethical problems that easily come with impressive technology, in the worst case leading to 'Disneyfication'. We will discuss this in some more detail below. Experience from museums suggests that "digital technologies work best when they enable people who feel connected to museum objects to have the freedom to deepen these relationships" (Newell 2012: 303). One way to do so is through using AR and place-based narratives in a playful, co-creative visitor experience.

3. Definitions

Before presenting our theoretical framework, we need to highlight some fundamental concepts. Being involved in any kind of capturing story – not the least in game play – often means experiencing a high level of immersion, a debated concept understood as the feeling of having left the ordinary world and entered that of the narrative (cf. Ryan 2001; Calleja 2010; Bowman 2018; Kingsepp/Van Den Heede 2022). Marie-Laure Ryan has suggested that virtual reality (VR) should be regarded as "a metaphor for total art": "[w]hat is at stake in the synthesis of immersion and interactivity is [...] nothing less than the participation of the whole of the individual in the artistic experience." (Ryan 2001: 20–21) In AR literature, immersion has been discussed through the sensation of being engaged and to perceive oneself as being part of the experience, then often in relation to AR games (Shin 2019). Fundamental to immersion is this sensation of full involvement in the constructed world. While focus here has generally been set on the computer-enabled overlay to form immersion (cf. Laffana et al. 2016; Morschheuser et al. 2017; Trunfio et al. 2022), in this article we want to highlight the role of physical elements for immersion in an AR setting. Physical elements are key to the experience of authenticity, a concept often used without definition. Authenticity is here understood as a feeling of something real and genuine, generated through "a style of visual representation whose goal is to make the viewer forget the presence of the medium [...] and believe that he is in the presence of the objects of representation" (Bolter/Grusin 1999: 272–73). Importantly, it is the viewer's intuitive assessment, based on previous experiences, which produces actual experiences of authenticity (ibid: 53).

4. Portals to time travel: From Heidegger's Greek temple to enchanted AR game play

Applying Lowenthal's idea to "fit a plug into the wall; and listen in to the past" (2015 [1985]: 55) to a present-day AR context implies that visitors at a heritage site will be able to have an immersive experience of not only listening to, but actually visiting the past, possibly using all of their senses while being there. They may even be able to interact with people they meet in this extra dimension. As a heritage site provides the narratives through which the past comes alive, and is in itself the magic portal through which it can be reached, we need to begin with the concept of place. Obviously, an augmented reality is ontologically close to the real world: what happens here takes place somewhere, and this is where we, inspired by Heideggerian phenomenology, begin the presentation of our model. Although place is central in Heidegger's philosophy, we will here refer to a section in his famous essay *The Origin of the Work of Art* (Heidegger 1987 [1935]). Here he uses a Greek temple as a site where the past and its people remain present, despite it not having been in religious use since antiquity. He regards the temple as situated in its natural surroundings, and notes that the spirit of the god, as well as the people whose lives were connected to the site in the past, are, somehow, still there. How is this possible? In Heidegger's view the world does not merely consist of material objects: instead, it is constantly being shaped into being through a network of relationships and meanings, a process he calls 'worlding'. Thus, the temple gathers elements from its surroundings and from the community of humans through which, and for whom, it was originally built. Situated in a historical context it is both part of the world *and* reveals a world, namely of its past (ibid: 38). The temple opens up a room in the world within which the life of the people can take place – which in our reading can be extended to not only a specific people, but also all who at some point visit the place. In particular, non-local visitors need to be acquainted with at least some of the historical narratives, or parts thereof – either before or during the visit – in order to experience the place from this perspective. However, as new narratives related to the site are continuously evolving, not the least through the visitors and their individual experiences, it is important for institutions to acknowledge this vitality (cf. Simon 2010; Refsland/Tuters/Cooley 2010).

5. Enchantment: Preparation for time-travel

Although Heidegger's influence on the way we understand the relations between human beings and their world has been widely acknowledged, not the least in Human Geography, it is another philosopher who has been our initial source of inspiration: Kendall J. Walton. Before presenting his theory about mimesis as make-believe (Walton 1990), we need to elaborate further on the implications of Heidegger's idea about the Greek temple as a room where history, nature and culture merge, enabling experiences of transcendence through 'opening up the world' (Heidegger 1987 [1935]). From another perspective, a seemingly related way to an experience of wholeness reaching beyond the everyday is through enchantment (or re-enchantment). Several thinkers from various disciplines have begun to question, and oppose, Max Weber's (1946 [1917]) influential idea about modernity as intrinsically disenchanted. While some argue that he was simply wrong, others propose that although the supernatural, magic, et cetera, is today widely associated with fairy tales and superstition, this is nevertheless something that fills an important function in contemporary society. Philosopher Jane Bennett belongs to the latter group of thinkers. Expanding the experience of enchantment to "the extraordinary that lives amid the familiar and the everyday" (Bennett 2001: 4), she also argues that enchantment is "an essential component of an ethical, ecologically aware life" (ibid: 99). Moreover, when immersed in the mood of enchantment, "we sense that 'we' are always mixed up with 'it', and 'it' shares in some of the agency we officially ascribe only to ourselves" (ibid: 99). This openness for the unexpected, and the acknowledgement of wholeness in what can be interpreted as a kind of collective transcendence, brings us to the closely related idea of 'as if', as presented by historian Michael Saler (2012).

6. 'As if': Entering history

In his book *As If: Modern Enchantment and the Literary Prehistory of Virtual Reality* (2012), Saler examines imaginary worlds related to fiction – including those of Sherlock Holmes, H. P. Lovecraft and J. R. R. Tolkien – that have become not only inhabited by their original characters, but also by millions of readers, viewers and fans. However, it would be erroneous to consider such worlds entirely based on fantasy and the unreal. On the contrary, as Saler argues,

"The vogue for fantastic imaginary worlds from the fin-de-siècle through the twentieth century is best explained in terms of a larger cultural project of the West: that of re-enchanting an allegedly disenchanted world. [...] Fantasy, cast in a rigorously logical mode replete with 'objective' details, was one solution to the crisis of modern disenchantment. The widely felt need of the period for forms of wonder and spirituality that accorded with reason and science helps us to understand why fantastic yet rational imaginary worlds proliferated at the turn of the century and after." (Saler 2012: 6–7)

Saler connects re-enchantment to what he calls public spheres of the imagination (ibid: 17–18) using the concept of 'as if': Through the combination of strictly rational form and imaginary, often fantastic narrative content, fictional worlds are collectively perceived 'as if' they were factually real. They are replete with objective details that enhance immersion and make it possible to experience a world where, for example, the supernatural is normal, and magic powers are something that can be learned, as in the Harry Potter stories. Accordingly, large numbers of people are today familiar with imaginary worlds founded on the playful oscillation between reality and 'as if'. The fascination for the fantastic and supernatural in contemporary popular culture is not restricted to works of fiction, but is found in for example reality-tv shows and documentary films, where participants go ghost hunting or learn about UFO incidents (Hill 2011). Whether this is best referred to in terms of re-enchantment, as proposed by Saler, or as a more sensitive mode of appreciation, as argued by, for example, Bennett, falls outside the scope of this article. Importantly, both strands of thought agree that life itself in modernity (or postmodernity) is filled with moments of enchantment. Bennett states that:

"[e]nchantment is something that we encounter, that hits us, but it is also a comportment that can be fostered through deliberate strategies. One of those strategies might be to give greater expression to the sense of play, another to hone sensory receptivity to the marvelous specificity of things." (Bennett 2001, 3-4)

The sudden experience and liminoid character of moments brought by enchantment seems comparable to Heidegger's concepts 'aletheia/Erschlossenheit' (disclosure), and 'Lichtung' (clearing), relating to a kind of transcendent experience of something beyond the everyday. It also links to the third philosopher in our model, Kendall J. Walton, who – as Bennett – regards surprise as a main link between the real world and imaginary ones, making the latter "come true" (Walton 1990: 14). We will now turn to his theory of mimesis as make-believe, which brings us closer to the vital role of physical objects in imagination and game-play.

7. Mimesis as make-believe: The role of the physical world in imagination

Although Walton does not refer to Heidegger, both focus on traditional fine art, in particular painting and sculpture, exploring their relations to the world, and how the spectator makes sense of them. For us, Walton offers the link to play and playfulness as part of the immersive experience at a site, and to the important role of physical, tangible objects in this. He delimits his theory not only to works of visual art, but also to fiction, which brings another link to the concept of 'as if'. However, he also opens the door to wider areas:

"We will be able to see representationality in the arts as continuous with other familiar human institutions and activities rather than something unique requiring its own special explanations." (Walton 1990: 7)

Walton (1990) regards most imaginings as "in one way or another dependent on or aimed at or anchored in the real world" (21). His concept fictional truth denotes that a proposition is valid only within the frames of a certain imaginary world: it is a fact that it is true in a game of make-believe. Importantly: when applied to a non-fictional setting such as a museum, a fictional truth would be a statement about the real world that includes a dimension of imagination and fantasy. This quality of 'as if' does not distort or transform facts to fiction, but rather adds a layer of playfulness and enchantment to rational thinking (Saler 2012).

Walton proposes that real things have three major roles in our imaginative experiences: they prompt imaginings; they are objects of imaginings; and as props they generate fictional truths. Prompters are founded in collective imaginative activities based on shared knowledge about, for example, a certain narrative or a fictional world familiar from popular culture (cf. Saler 2012; Jenkins 2006). This means both that the imaginings can be spontaneous, and that they can – at least to some extent – be predicted. Moreover, "it is probably obvious to each participant that the others will imagine what he does" (Walton 1990: 23).

This can be expanded to situations where the digital and the physical are combined, for example a heritage site, especially when this can provide feelings of enchantment. In this context, the concept may be understood in terms of a sudden transcendence, as well as an adventure characterised by 'as if'. Most probably, one does not exclude the other. In this scenario, then, the props would generate not fictional, but enchanted truths: in other words, the statement is true not only in a game of make-believe, but also in the real world at the site. What has happened is that something actual has suddenly become actual and extraordinary: an object or idea usually assessed with rational reason – thus in a way disenchanted – is suddenly experienced from a non-rational, playful perspective that cognitively transforms it into a fascinating hybrid form. In fact, Walton (1990) indicates support for this idea, as "[w]hat is fictional need not be imagined, and perhaps what is imagined need not be ficticnal." (37) Thus, what is imagined can be based on an augmented reality provided by digital layers are added to the physical space.

8. Theorising re-enchantment: AR solutions at the Långban heritage site

We will now apply the theoretical framework on three AR prototypes designed for the heritage site and open-air museum at Långban, Värmland county, Sweden. Founded in the early 16th century, Långban used to be a vibrant mining village, but was transformed into a museum in the 1970s, when the mine was closed down. Today, visitors can enter the remaining buildings hosting exhibitions of the mine's history, and a collection of its rare minerals. The mine itself cannot be entered, as it has been filled with water, but visitors can take walks in the surroundings, today a nature reserve with unique fauna due to the mineral rich soil, or search for their own pieces of mineral in the quarry.

The Långban museum's task, being a state and regionally funded institution, is on the one hand to preserve Sweden's industrial heritage, and on the other to communicate the past to the present. Alike many museums and heritage sites they want to engage visitors with history and make it accessible by developing AR solutions. A carefully outlined participatory design process was conducted in collaboration with a wide array of actors. There were experts on the history of the site, visitors, artists and media experts with extended knowledge in digital communication, stakeholders from different local authorities and researchers. Using participatory design allowed for a democratic, inclusive learning and development process (Bjögvinsson et al. 2012) and to ensure that the specific site stayed in focus (the participatory design process is outlined in detail in Ryan Bengtsson et al. 2022). Through linking the real world and the present Långban with the past Långban, an imaginary bridge can invite the visitors to physically experience entering an inaccessible mine, and meeting inhabitants from times past.

The first AR installation we will discuss, the *Chtonoscope*, is rather a *Chronoscope*, a device to travel back in time through oscillating between visuals from the present and the past. The second installation, the *Letter Box*, uses personas of past inhabitants to form a dramatized digital narrative where an AR version of her/him guides the visitor through the landscape. The third example, the *Echo Capsule*, allows visitors to explore the depth of the mine using their own voice and to leave messages for friends and future visitors, thereby creating links from the present to what in the future will become the past.

Before going into the specifics of each prototype, we must begin with the place itself. As these digital installations are located in the landscape at the heritage site, the locale contributes significantly to the setting, or the stage, of the experience. For the visitor, being there physically is the first symbolical step towards place-based time travel. When the digital content merges with the physical setting through AR, there is a "remediation of place" (cf. Bolter/Grusin 1999; Jaworski/ Thurlow 2011; Ryan Bengtsson 2012; McQuire 2016). Following Walton's theory, the work can also be carried out in the visitor's imagination, as long as suitable prompters/props have been provided. The presence of additional digital layers at the place together with real, tangible objects would contribute to the experienced authenticity of the invoked narrative unfolding in the visitor's own mind (cf. Haraway 1997). The physical setting anchors the narrative in the real world and provides its framework.

9. Plugs into the wall

In order to spark the individual visitor's imagination into a game of make-believe, there needs to be a shared agreement that a particular installation functions as a prompter; that it, so to speak, gives a signal that a particular game begins. While the device itself might either look familiar (the *Letter Box*) or more or less alien (in particular the *Chronoscope*), it is necessary that its physical presence at the location indicated that engagement will result in something potentially surprising – in other words, that the visitor will identify it as a prompter, as compared to the ordinary 'pull the handle and see the text in the box' experience. How can this be achieved?

One way is to prepare the visitor, so that she or he is already disposed towards playfulness when finding the object. On arrival at Långban, the visitors are encouraged to download an App designed to function as a magic portal to invisible dimensions at the site through AR. While those who prefer other forms of guidance may still enjoy their visit, those who choose to use the app will be able to take advantage of its options for co-creation. When this visitor is invited to open the *Letter Box* lid, place her/his hands on the *Chronoscope*, and use the microphone at the podium of the *Echo Capsule*, it is highly probable that the prompter character of these props is recognised. As compared to the purely causal function of ordinary lids and optical devices, prompters invite curiosity and ask to be handled, which believably will lead the visitor further into a mood of playful immersion. When used in a context of historical narratives at the site, the props serve as the devices through which the visitor navigates her/his time machine.

In the next section we will present the three AR installations at Långban in more detail, and relate them to our theoretical framework.

9.1. Prompters and props for enchantment, # 1: The Chronoscope

The *Chronoscope* – which also works as a Chtonoscope – is designed to evoke the visitors' curiosity by sticking up from the ground, thereby attracting attention. It looks similar to a periscope: a fascinating object, well known from popular culture, the use of which is not familiar to everyone. Regarded as a prop, the *Chronoscope* indicates something magical, enchanted, and extraordinary. The handles intrigue the visitor to grab them and look into the device, which now turns into a prompter: not only does the deep below the earth's surface become visible, which in itself is 'magic'. On top of that, it also turns out to be a virtual wormhole to the past.

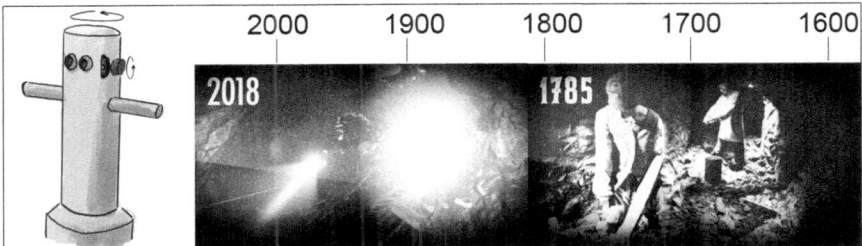

Fig. 1: Illustration of the Chronoscope by © Sticky Beat. (Screenshot from Sticky Beat 2020, Url: https://stickybeat.se)

The handles suggest that you can turn the *Chronoscope*, and when doing so, the perspective shifts, as well as time. Thus, the handles are a time control, allowing the user to alter between different periods and activities in the mine. The visitor navigates across a digital history line of images, drawings and short film clips, ranging from the 16th century mine workers to the contemporary water-filled mine with mine-divers swimming by. The physical movement and visual timeline open for playful and explorative interaction with the hitherto unknown subterranean territory, both in its present state below one's feet and in its revived past. Of course, the visitor is well aware that we do not actually look into a wormhole. However, in this setting we may collectively accept this as an 'as if' experience, thereby adding a dimension of authenticity to our experience.

9.2. Prompters and props for enchantment, # 2: The Letter Box

The *Letter Box* is an actual letterbox, but when opening the lid, the visitor finds a letter from someone living at Långban in the past. As a letterbox is a familiar object that we interact with in our daily lives, its inspirational potential would possibly be limited. Nevertheless, its location in this particular setting suggests that there will be something in it going beyond the everyday. Perceived as a prop, the letterbox becomes a promise of enchanted communication.

The letter consists of an image, or card, representing a historical persona. It contains a bio of her/him, a description of a dilemma s/he struggles with, and an invitation to walk together through the heritage site. You accept the invitation through scanning a QR code. Each persona is of a different gender, age, has different occupations, and comes from different periods in time. While the facts are based on archival material from the village, the narratives are also inspired by other writings and literature referring to the specific periods, partly constructing them as fictional truths.

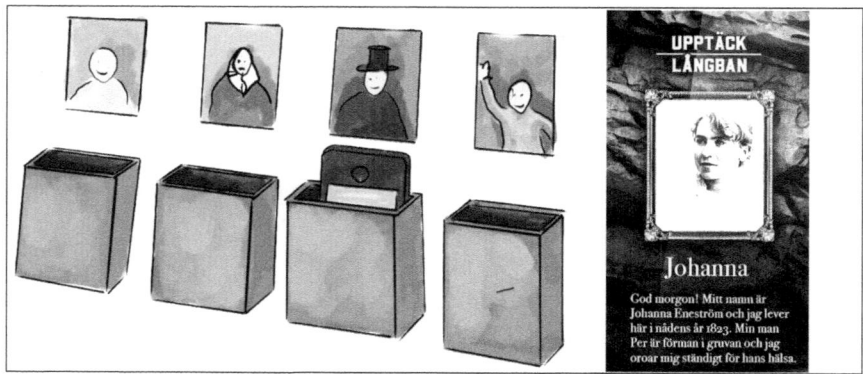

Fig. 2: Illustration of the Letter Box by © Sticky Beat. (Screenshot from Sticky Beat 2020, Url: https://stickybeat.se)

The AR versions of the personas tell the stories of their lives from their perspectives, including references to sites in the village that are of importance to them. These sites set the scene for the narrative, as for example when visiting the school together with a young girl from the 17th century. The AR application overlays the sites with a visualisation, or scene, adapted from the life of the persona, thereby inviting the visitor to engage in her/his everyday life. Architecture and specific marks in the landscape introduce new parts of the narrative, mapping a digital and physical path through the site. Using specific sites in the landscape as props, the narrative unfolds across the village in unforeseen ways, physically and imaginatively involving the visitor in this process.

At certain points, the visitors and the apparition with whom they walk around might meet another visitor, who is temporarily following alongside the paths of another historical character. At these encounters, the geographical point interconnects different timelines, different personas and different visitors. These crossings enable a dialogue across time boundaries as, potentially, time evaporates. When visiting with friends or families, each visitor selects her/his own friend from the past, which means that when comparing their experiences, they all have the opportunity to participate in the same game of make-believe, albeit through different perspectives. The intersecting narratives open for further play and/or discussion, either at the site, or after the experience.

9.3. Prompters and props for enchantment, # 3: *The Echo Capsule*

The *Echo Capsule* consists of a podium placed next to the mine's opening. The visitor is invited to place her/his mobile phone on the pedestal, turning the combined props into a prompter. While doing so, s/he will notice that the mobile can be tilted in different positions. Below, there is a microphone, another familiar object that here evokes expectations and curiosity about what will happen when you make a sound

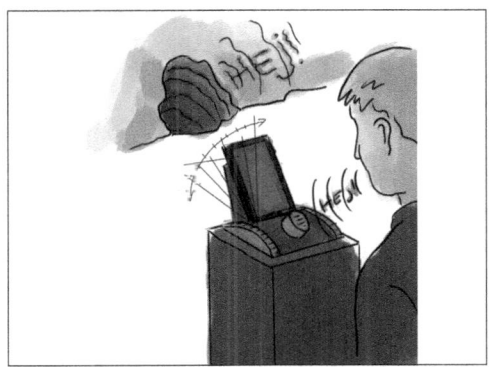

Fig. 3: Illustration of the Echo Capsule by ©
Sticky Beat. (Screenshot from Sticky Beat 2020.
Url: https://stickybeat.se)

The sound turns out to get repeated through an echo from the depths of the shaft, illustrating the time it takes for the sound to travel into the mine and back again. By tilting the mobile, you can define how deep into the mine you want to direct the sound, which affects the echo's delay. The tilting mechanism informs you of meters, as well as the year the miners reached this depth/ level. The visitor may also choose to record a message, and define when it is to be repeated, ranging from a few seconds later to days ahead. S/he can then notify a friend about being at this place at a specific time, in order to receive a surprise. Here, playfulness is at the core of the experience, as the combination of tactile and auditory experiences enables the user to play with both spatial and temporal distance. The *Echo Capsule* is a device which could probably be experienced as rather meaningless, unless the visitor has initially agreed to adopt a mode of 'as if', thereby being receptive to invitations to play.

To summarise: the site first needs to be established as enchanted, in this case primarily understood as a place for playfulness and time travel, while also enabling possible experiences of enchantment on a more profound level. This may be achieved through AR by for example an adequately designed mobile phone app. When having accepted the invitation to play, the visitor enters an extraordinary dimension of the site where past, present, and future turns out to have merged. Here the visitor has the ability to identify certain objects as being potentially marvellous when handled (props that become prompters), leading to co-creative narrative experiences. The combination of the physical, tactile, and emotional, all taking place within an imagined framework of an enchanted site, creates a valuable foundation for an immersive experience that also includes informative and intellectual dimensions.

10. Time travel and other magical experiences

When discussing contemporary use of digital media in museums and heritage settings it is common to hear critical comments about the way some museums present information in a semi-interactive way: by simply letting the visitor open boxes, pressing buttons, et cetera, written text and/or (moving) images appear. Indeed, it is a challenge integrate media technology in ways that create engaging experiences in this type of context (Drotner et al. 2019). However, it is important to move beyond the possible disappointment occurring when met by something familiar, thus not very exciting, or something that does not live up to the promise of being potentially surprising. Importantly, we do not argue that a visit to a museum should be characterised by spectacle. Instead, this is where the theories about enchantment become fruitful, allowing for thoughtful discussion and development of AR installations in open-air museums. The AR examples presented above use playfulness, inviting the user to play through well designed props and prompts that utilise the landscape and the visitors' imagination. Playfulness allows for enchantment and immersion in the narrative, if not entirely, then at least partially. We therefore suggest that these concepts contribute to further understanding of how a historical site, physical objects and digital technologies can interplay for engagement. Further, we do not suggest that AR is a superior way of presenting historical narratives, or engaging people in a dialogue with the past; rather, it should be regarded as an addition to other techniques, another way of 'how to do things with history'.

We here propose a carefully considered conceptual setup that may serve as a framework for further elaborations when/if put into practice. In a museum setting it is important to enable and discuss aspects of using games and gamification in historical contexts, not only regarding how experiences are constructed but, more importantly, how can AR (and other digital solutions) allow for a variety of engagements. We believe these should include joy and learning as well as critical thinking and reflections on both the past and the present. In the ongoing digitalization of heritage, there is a need for well-founded theoretical models that can enable substantiated discussions about it.

The Långban examples illustrate how AR technology may provide a major contribution to invoking a mood of enchantment that will invite visitors to play with time, place, and history (cf. Bennett 2010). Combining the narrative framework provided by a geographical location/place, digital media, material objects (props), and the advantages brought by collective imagining, may potentially result in not only a higher level of immersion, but also a more memorable experience at the site (cf. Marchant/Nancarrow 2019). While the term immersion has been widely used for the all-encompassing experience of disappearing into a written text, it has also been challenged, in particular within a digital games' context (see Calleja 2010; Bowman 2018). However, as the AR devices discussed here are relying heavily on the user's own imagination in response to the presented narratives, Ryan's (2001) perspective on immersion is still fruitful.

Digital technologies can enhance and extend heritage, adding layers of voices, narratives and meanings that are otherwise hard to engage with (cf. Newell 2012). In order to communicate with its audiences, offer new and additional perspectives and new understandings, and interconnecting the past with the present excitement and emotional engagement with the museum's content are valuable assets, especially regarding visitors with little or no pre-knowledge (Cameron/Kenderdine 2010). Importantly, digital media technology does not facilitate immersion by itself. AR – like any digital solution – should involve the visitor not merely as someone pressing a button or moving a lever. Critics may express concern about the level of accuracy, or 'truth', in the representations being diminished if engaging experiences are added. However, visitors do understand that what is playfully imagined is not necessarily in accordance with historical reality, even when they are entwined (Saler 2012). By crafting a 'magic wormhole' the place itself, its history and existing narrative become part of a dimension of fictional/enchanted truth, to which the visitor is invited through adopting an 'as if' mode. Here, we understand museum visitors as potential participants in the narratives, oscillating between fact, fiction, and imagination.

The use of play and imagination in a heritage context can attract visitors other than those already interested in history, or familiar with visiting museums, thereby becoming more inclusive (cf. Newell 2012). It is crucial to acknowledge the difference between the experiences we have referred to here and experiences built on simulacra, thus without any relation whatsoever to the real world (Baudrillard 1994 [1981]). Authenticity in the Långban examples is not based on purely technological means, as the site itself provides the bedrock for visitor engagement. This important physical dimension anchors the AR experiences and turns them into not just advanced plugs into the wall: they are designed to be short, temporary wormholes in time that enable enchanted experiences. The games of make-believe at Långban take place within a narrative framework provided by the site's history, including real people who are represented as individuals who lived and worked there. To play with them is also to remember them, and to get closer to their life stories. In this way, playfulness is ethics in practice.

Acknowledgements

This work was supported by BFUF (the R&D Fund of the Swedish Tourism and Hospitality Industry) and the European Regional Development Fund under Grant 20201439.

References

Baudrillard, J. (1994 [1981]): Simulacra and Simulation. Ann Arbor: University of Michigan Press.
Bennett, J. (2001): The Enchantment of Modern Life: Attachments, Crossings, and Ethics. Princeton: Princeton University Press.
Björgvinsson, E./Ehn, P./Hillgren, P.-A. (2012): "Design Things and Design Thinking: Contemporary Participatory Design Challenges." Design Issues 28(3), pp.101–116.
Bolter, J./Grusin, R. (1999): Remediation: Understanding New Media. Cambridge, Mass.: The MIT Press.
Bowman, S. L. (2018): "Immersion and Shared Imagination in Role-Playing Games." In: J. P. Zagal/S. Deterding (eds.), Role-Playing Game Studies: Transmedia Foundations. New York: Routledge, pp. 379–394.
Braunerhielm, L./Ryan Bengtsson, L. (2023): "Geomedia Sensibility in Media Technologies." Anatolia 35(3), pp. 506–516.
Braunerhielm, L./Gibson, L./Ryan Bengtsson, L. (2024): "Geomedia Perspectives for Multiple Futures in Tourism Development". Media and Communication 12(8157), 1–16.
Bujari, A./Ciman, M./Gaggi, O./Palazzi, C. E. (2016): "Using Gamification to Discover Cultural Heritage Locations from Geo-tagged photos." Personal and Ubiquitous Computing 21(2), pp. 235–252.
Caillois, R. (1958): Les Jeux et les Hommes. Paris: Librairie Gallimard.
Calleja, G. (2011): In-Game: From Immersion to Incorporation. Cambridge: MIT Press.
Champion, E./Bharat D. (2010): "Dialing up the Past." In: F. Cameron/S. Kenderdine (eds.), Theorizing Digital Cultural Heritage: A Critical Discourse. Cambridge: MIT Press, pp. 333–347.
Chiu, C. K./Tseng J. C. R./Hsu, T. Y. (2017): "Blended Context-aware Ubiquitous Learning in Museums: Environment, Navigation Support and System Development." Personal and Ubiquitous Computing 21(2), pp. 355–369.
Couldry, N./ McCarthy, A. (2004): Mediaspace: Place, Scale and Culture in a Media Age. London: Routledge.
De Souza e Silva, A. (2006): "From Cyber to Hybrid: Mobile Technologies as Interfaces of Hybrid Spaces." Space and Culture 9(3), pp. 261–278.
Fast, K./Jansson, A./Tesfahuney, M./Ryan Bengtsson, L./Lindell, J. (2018): "Introduction to Geomedia Studies." In: K. Fast/A. Jansson/M. Tesfahuney/L. Ryan Bengtsson/J. Lindell (eds.), Geomedia Studies: Spaces and Mobilities in Mediatized Worlds. London: Routledge, pp. 1–17.
Flynn, B. (2010): "The Morphology of Space in Virtual Heritage." In: F. Cameron/S. Kenderdine (eds.), Theorizing Digital Cultural Heritage: A Critical Discourse. Cambridge: MIT Press, pp. 349–368.

Heidegger, M. (1987 [1935]): Konstverkets ursprung [Der Ursprung des Kunstwerkes]. Göteborg: Daidalos.
Hill, A. (2011): Paranormal Media: Audiences, Spirits and Magic in Popular Culture. London: Routledge.
Hjorth, L./de Souza e Silva, A. (2023): "Playing with Place: Location-based Mobile Games in Post-pandemic Public Spaces." Mobile Media & Communication 11(1), pp. 52–58.
Huizinga, J. (1938): Homo Ludens: Proeve eener Bepaling van het Spel-element der Cultuur. Haarlem: Tjeenk Willink.
Javornik, A./Kostopoulou, E./Rogers, Y./Fatah gen Schieck, A./ Koutsolampros, P../Moutinho, A. M./Simon J. (2019): "An Experimental Study on the Role of Augmented Reality Content Type in an Outdoor Site Exploration." Behaviour and Information Technology 38(1), pp. 9–27.
Jaworski, A./Thurlow, C. (2011): "Tracing Place, Locating Self: Embodiment and Remediation in/of Tourist Spaces." Visual Communication 10(3), pp. 349–366.
Jenkins, H. (2006): Convergence Culture: Where Old and New Media Collide. New York: New York University Press.
Jordanova, L. (2006): History in Practice. 2nd ed. London: Bloomsbury Academic.
Kingsepp, E. (2006): "Immersive Historicity in World War II Digital Games." Human IT: Journal for Information Technology Studies as a Human Science 8(2), pp. 60–89.
Kingsepp, E. (2018): "Experiencing and Performing Memory: World War II Videogames as a Practice of Remembrance." In: P. Finney (eds.), Remembering the Second World War. London: Routledge, pp. 217–233.
Kingsepp, E./Heede, P. V. (2022): "Immediacy/Hypermediacy." In: Bloomsbury History: Theory and Method. London: Bloomsbury.
Laffana, D. A./Greaneya, J./Bartona, H./ Kaye, K. L. (2016): "The Relationships Between the Structural Video Game Characteristics, Video Game Engagement and Happiness among Individuals who Play Video Games." Computers in Human Behavior 65, pp. 544–549.
Laine, T. H./Suk, H. J. (2016): "Designing Mobile Augmented Reality Exergames." Games and Culture 11(5), pp. 548–580.
Landy, J./Saler, M. (2009): The Re-Enchantment of the World: Secular Magic in a Rational Age. Stanford: Stanford University Press.
Lefebvre, H. (1991): The Production of Space. Malden: Blackwell.
Lowenthal, D. (2015 [1985]): The Past Is a Foreign Country: Revisited. Cambridge: Cambridge University Press.
Marchant, A./Nancarrow, J. H. (2019): "Introduction: Practice, Performance, and Emotions in Medieval and Early Modern Heritage." Parergon 36(2), pp. 1–16.
McQuire, S. (2016): Geomedia: Networked Cities and the Future of Public Space. Cambridge: John Wiley and Sons.
Morschheuser, B./Riar, M./Hamari, J/Maedche, A. (2017): "How Games Induce Cooperation? A Study on the Relationship between Game Features and We-

Intentions in an Augmented Reality Game." Computers in Human Behavior 77, pp. 169–183.

Murray, J. H. (1997): Hamlet on the Holodeck: The Future of Narrative in Cyberspace. Cambridge: MIT Press.

Nancarrow, J. H. (2019): "Hidden Heritage: Concealment, Reuse, and Affective Performance in Historic Buildings and Digital Heritage." Parergon 36(2), pp. 63–89.

Newell, J. (2012): "Old Objects, New Media: Historical Collections, Digitization and Affect." Journal of Material Culture 17(3), pp. 287–306.

Parry, R. (2010): "The Practice of Digital Heritage and the Heritage of Digital Practice." In: R. Parry (eds.), Museums in a Digital Age. London and New York: Routledge, pp 1–9.

Pierroux, P./Bäckström, M./Brenna, B./Gowlland, G./ Ween, G. (2020): "Museums as Sites of Participatory Democracy and Design." In: P. Pierreoux/P. Hetland/L. Esborg, A History of Participation in Museums and Archives. Routledge: London, pp. 27–45.

Refsland, S./Tuters, M./Cooley, J. (2010): "Geo-Storytelling: A Living Archive of Spatial Culture." In: F. Cameron/S. Kenderdine (eds.), Theorizing Digital Cultural Heritage: A Critical Discourse. Cambridge: MIT Press, pp. 409–416.

Ryan, M-L. (2001): Narrative as Virtual Reality: Immersion and Interactivity in Literature and Electronic Media. Parallax: Johns Hopkins University Press.

Ryan Bengtsson, L. (2012): Re-negotiating Social Space: Public Art Installations and Interactive Experience. PhD dissertation, Karlstad University.

Ryan Bengtsson, L./Braunerhielm, L./Gibson L./Hoppstadius, F./Kingsepp, E. (2022): "Digital Media Innovations through Participatory Action Research: Interventions for Digital Placebased Experiences." Nordicom Review 43(2), pp. 134–151.

Saler, M. (2012): As If: Modern Enchantment and the Literary Prehistory of Virtual Reality. Oxford: Oxford University Press.

Samuel, R. (2012): Theatres of Memory: Past and Present in Contemporary Culture. Revised ed. London: Verso.

Shin, D. (2019): "How does Immersion Work in Augmented Reality Games? A User-Centric View of Immersion and Engagement." Information, Communication and Society 22(9), pp. 1212–1229.

Simon, N. (2010): The Participatory Museum. Santa Cruz: Museum 2.0.

Sturani, M. L. (2022): Landscape as Heritage in Museums: A critical Appraisal of Past and Present Experiences: Landscape as Heritage. London: Routledge, pp. 289–300.

Thielmann, T. (2010): "Locative Media and Mediated Localities." Aether: The Journal of Media Geography 5(1), pp. 1–17.

Trunfio, M./Lucia, M. D./Campana, S./Magnelli, A. (2022): "Innovating the Cultural Heritage Museum Service Model through Virtual Reality and Augmented Reality: The Effects on the Overall Visitor Experience and Satisfaction." Journal of Heritage Tourism 17(1), pp. 1–19.

Uricchio, W. (2011): "Simulation, History, and Computer Games." In: J. Raessens/J. Goldstein (eds.), Handbook of Computer Game Studies. Cambridge: MIT Press, pp. 327–338.

Walton, K. L. (1990): Mimesis as Make-Believe: On the Foundations of the Representational Arts. Cambridge: Harvard University Press.

Weber, M. (1946 [1917]): "Science as a Vocation." In: H.H. Gerth/C. Wright Mills (eds.), From Max Weber: Essays in Sociology. New York: Oxford University Press, pp. 129–156.

Culture and Emotions as Historical Game Design Affordances

Cultural Combinatorics and Conjured Spectres
The Representation of Culture and Cultural Hybridity through the Game Mechanics of *Crusader Kings III*

Michael A. Conrad

Abstract

Crusader Kings III is a popular medieval-themed, grand strategy game that features a complex culture system. With the Royal Court DLC, new mechanics were introduced to simulate cultural encounters that allow players to create hybrid cultures. The article examines the workings of these game mechanics, their strategic purposes as well as the ensuing concept of culture by applying different methods, including a quantitative analysis of cultures in the game data for identifying possible design biases and how strategic advantages might stem from the distribution of cultures at game start. By analysing the game's structures and the rendering of cultural processes, the theoretical framework guiding the creative design is explored as well. Ultimately, the study seeks to better understand how historical games foster engaging relationships with the past through gameplay.

Keywords

historical games, game mechanics, digital humanities, cultural hybridity, medievalism

1. A hauntological opening

"So it would be necessary to learn spirits. [...] To live otherwise, and better. No, not better, but more justly. But *with them*. [...] And this being-with spectres would also be, not only but also, a *politics* of memory, of inheritance, and of generations." (Derrida 1994: xviii–xix)

The past is gone; there is no returning to it. Yet, it lingers with us, haunting our present through the traces of past decisions, events, customs, and traditions. In this sense, the past is both *with* us and *beyond* our reach – a *spectre* that is neither fully absent nor fully present, that links past and present (Buse/Scott 1999: 11). Consequently, culture always carries a spectral quality, making its analysis a 'hauntological' project, one that deals with what does not fully exist but still shapes our everyday lives (cf. Fisher 2013). In this way, culture responds to the human desire to forge emotionally and intellectually meaningful connections with the past, using methods that summon its spectres, granting them a sensory form to bring them back into the heart of our communities.

In this context, historical storytelling can serve as a conjuring ritual. While such entails an intellectual endeavour mostly, play offers a more sensory-driven alternative. Historical reenactments, for instance, aim to create immersive experiences of historicity through role-play. Three key factors distinguish playful engagement with history from academic approaches: immersion, imitation (or, according to Roger Caillois, *mimicry*), and a fixed set of rules (Huizinga 2002 [1938]; Caillois 2001 [1958]). Imitation enhances immersion by minimising the material differences between game worlds and their historical counterparts. Less obviously, rules are not only essential for structuring a game world – they also link past and present. In fact, defining rules is itself an act of the past that has consequences for the present and therefore necessitates a spectral quality. Consider modern chess: its rules, codified in the 15th century, remain in use even today. Each time we play, we re-enact and re-actualise them as historical artifacts.

On the other hand, historical computer games share many similarities with historical reenactments, as evidenced by Carl Heinze's (2012: 85) definition: "A computer game is a historical computer game if a functionally relevant part of its ludic elements refers to meaning that is constructed by historical discourse and that belongs to collective historical knowledge."[1] This relationship between games and historical discourse forms the basis for the following analysis of how *Crusader Kings III* (Paradox 2020; hereafter CK3) represents culture. Rather than focusing on the game's rich visual portrayal of culture and its potential to reinforce historical biases, however, this discussion centres on culture as a *game mechanic*. Game mechanics – like audiovisual elements – shape a game's storytelling, though more subtly, primarily through abstract structures rather than direct sensory input. As such, mechanics are no less responsible for how players may experience spectres of the past in the present.

In this regard, CK3 is particularly compelling due to its intricate culture system. Analysing this system allows for a deeper understanding of some of the underlying notions of culture embedded within its mechanics, as well as of the cultural theories that may have informed their design. The key example for exploring these theoretical aspects is cultural hybridisation, which is a mechanic that was added to the game at a later stage and will serve as a central focus of this article. For more than many others, this mechanic illustrates how much CK3's approach to culture differs from others in its aspiration to offer players a more realistic, multi-dimensional, experience of past cultures – as a dynamic element that both orients and motivates gameplay. As such, it remains ambivalent: on

---

[1] My translation.

the one hand, like cultural identity in real life, the mechanic constrains counterfactual storytelling by anchoring it in seemingly authentic models that align with players' expectations and their knowledge of history. On the other hand, it allows for greater flexibility by enabling players to form alternative narratives and identities. In doing so, it offers a more nuanced portrayal of past cultures and their dynamics, including the role cultures play as significant drivers of historical change.

The specific complexities of the employment of culture in CK3 warrant a similarly nuanced analytical approach. For that reason, a textual analysis of a multitude of sources is employed, including such documents as the official CK3 wiki (i.e., the *CK3 Community Wiki*), developer diaries and relevant community forums. In addition, preliminary results from a quantitative analysis of game data support the other observations on CK3's culture system. They shed light on the distribution of cultures and culture groups at a specific starting point (the year 1066), helping to identify potential design biases that could create cultural dominance or strategic advantages for certain cultures. These findings are further contextualised through personal gameplay experience, adding a reflexive dimension to the analysis and supporting a more nuanced interpretation of the game's cultural logic.

2. Accuracy vs. authenticity

Can media be a means to bring "spectres" of the past back home, to conjure and thereby re-actualise them intentionally in our shared present? How much of its bygone reality can we ever hope to meaningfully revive in order to approximate what it might have felt like to 'live back then'? Whenever media engage with the past, questions of historical authenticity arise almost immediately and somewhat instinctively. Although the debate on historical authenticity is longstanding, winding, and unresolved, it cannot be ignored. Any serious attempt to understand and evaluate the representation of past cultures in computer games must presuppose the existence of reliable historical knowledge about how things were (or at least *most likely* were), as well as scholarly methods for establishing such knowledge. Recent academic discourse on the authenticity in historical reenactments can be a good starting point, since they offer a model for also discussing authenticity in computer games.

Whereas older strands of the debate focussed on linking authenticity to objectivity, current authors generally agree that authenticity has more to do with emotions than historic evidence. Dutch philosopher Robbert-Jan Addriaansen (2022: 54), for instance, starts his text on historical reenactments by discussing the phenomenon of "period rush", that is, the thrill of being immersed in the past when reenacting a historical event. As he further explains, this emotion conveys what Johan Huizinga (2014: 54) calls "historical sensations", the feeling of a "not completely reduceable contact with the past" in terms of "an entry into an atmosphere" that "presents itself in a single moment" (ibid.: 54). However, this intuitive connection is not identical with an experience of past itself. Instead, reenactors are "immersed [...] in the play of the reenactment" (Carretero/Wagoner/Perez-Manjarrez 2022: 11). What this implies is that historical simulations, including games, do not represent the past as such but offer models upon which *interpretations* rooted in collective historical knowledge are constructed. In his analysis of computer games set in historic periods, such as *Assassin's Creed*, Carl Heinze (2014: 179–183) confirms that what is experienced through such gameplay is a fiction that evokes *feelings* of authenticity.

That said, past debates on authenticity have led to rather fruitless discussions, as they ignore the underlying "clash of object and subject authenticity" (Zimmermann 2020: 15). Therefore, a more useful approach is to distinguish between *authenticity* and *accuracy*. Historicity and accuracy align more closely with 'object authenticity', as both rely on verification – such as confirming dates, agents, and artifacts. However, due to the nature of digital games and their virtual worlds, they cannot be "verified in the same way a historical source could be" (ibid.). In this sense, an emphasis on accuracy may overcompensate for what the medium lacks in terms of reality status. What game scholars should focus on instead is 'subjective authenticity' as the result of "convincing atmospheres" (Zimmermann 2020: 15–17). As much as that may be the case, one should not overlook that there nonetheless is a strong correlation between accuracy and *persuasion*: the more historical details are present within a virtual world, the more convincing it may become in fostering "suspension of disbelief" (Brown 2012). This is mirrored in the "great efforts" that the developers put into CK3's artwork:

"We made a great effort to keep a good level of historical accuracy in our designs. Our illustrations and icons are made to reflect the time period, just like the 3D art that populate the map. Of course, sometimes we had to try and find good compromises in the designs that would work for the entire timespan of the game. Where possible, we based clothes on reconstructed sewing patterns from extant medieval clothing (CK3 Dev Diary #28)."

The unequally distributed ability to recognise historical accuracy can, however, create divisions among player communities, separating those who are historically literate and capable of critically engaging with historical details from those who are not. In an era marked by widespread disinformation, it furthermore seems unwise to embrace epistemic relativism fully by abandoning scientific criteria for distinguishing subjective beliefs from objective facts. While authenticity may be a subjective experience, historical accuracy must still be assessed through comparisons between historical knowledge as *comparandum* and its representation as *comparans*.

3. Counterfactual storytelling in CK3

CK3 is the successor to the highly successful *Crusader Kings II* (Paradox 2012) and remains one of the most complex medieval-themed historical games designed for mass audiences. Prior to the game's official release, Paradox began publishing the first entry for its series of developer diaries on 24 October 2019, outlining core concepts and major changes. Intended to accompany the production process, these diaries serve as a platform for the communication between designers and the player community. Beyond merely disseminating information, the diaries also function as a marketing tool, which helps explain their occasionally colloquial tone. Nevertheless, they remain a valuable source, offering unique insights into the game's development and providing indications of the design principles guiding its creation. Publishing under the pseudonym of 'Doomdark', in the first diary the Chief Creative Officer explains four key points of the creative concept:

"Character Focus: Crusader Kings is clearly and unequivocally about individual characters, unlike our other games. This makes CK most suited for memorable emergent stories, and we wanted to bring characters into all important gameplay mechanics (where possible).
Player Freedom and Progression: We want to cater to all player fantasies we can reasonably accommodate, allowing players to shape their ruler, heirs, dynasty and even religion to their liking – though there should of course be appropriate challenges to overcome.
Player Stories: All events and scripted content should feel *relevant, impactful* and *immersive* in relation to the underlying simulation. That way, players will perceive and remember stories – their own stories, not the developers' stories.
Approachability: Crusader Kings III should be user friendly without compromising its general level of complexity and historical flavor. It's nice if it's easier to get into, but more than that, it should be clear what everything in the game is, what you might want to be doing, and how to go about it. (CK3 Dev Diary #0)"

What becomes apparent here is the extent to which the game was reconstructed to grant players as much autonomy as possible in shaping their own stories by incorporating more elements from role-playing games (RPGs). This is reaffirmed in a diary entry for the console version (15 March 2022), where the game is characterised as a 'grand strategy game with an RPG twist' and a 'sandbox nature', with the overarching objective being "the continuation of a Dynasty through the ages" (CK3 Console Dev Diary #2, n. p.). Indeed, mechanics such as stress management, character traits and skills were added, all of which have a decisive impact on what players can or cannot do. Several DLCs (downloadable content) published in the following years added further specialised events and mechanics to the game, thereby enhancing the game's level of detail. The aforementioned 'sandbox nature', however, has been a defining feature of the game from its earliest days. Instead of telling 'the developers' stories', as is common for many narration-centred games, players can use their resources, skills and traits for setting their own strategic goals, for shaping emergent narratives of their own. This idea is well-encapsulated in one of Paradox's slogans: "We make the games, you create the stories." (CK3 Dev Diary #22, n. p.)

While the CK series is known for its passion for historic detail, one adjective dropped in the first entry of the developer diary is quite revealing: "flavo[u]r", as it indicates that the design goal of this obsession with accuracy is not so much to create faithful portrayals of history but to provide a backdrop for medievalist phantasies. Therefore, what seems even more important than building perfect illusions of *objective* authenticity is to avoid anachronisms, as they could potentially break the illusion of *subjective* authenticity (cf. Pitruzzello 2014: 49). It is this emotional quality of *felt* authenticity that is crucial for the players in their desire to *feel* as immersed in the game world as possible, since this lays the groundwork for motivating them to tell their own stories (mostly) to themselves. That way, CK3 provides a simulative model based on a reference system informed by historical knowledge (cf. Heinze 2012: 99), with fragments of this knowledge serving as narrative material (cf. Frasca 2003; Juul 2005: 156–159). Although storytelling in CK3 is, naturally, limited to the elements the game provides, the great player autonomy makes it less deterministic and more aligned with an explorative playing style. In simple terms, *storytelling in CK3 is counterfactual – yet still guided by historical evidence – while being self-directed and driven by authoritative decisions*. In their gameplay, players explore alternative histories that diverge from historical starting points set by the game system but are based on real historic situations. Or, in other words, the game incorporates history as "recognizable history" (Aust 2009: 159). For example, at the beginning of the game, players can pick from a variety of historic characters. From that point onwards, though, the alternative storyline, driven by player decisions, will branch off, so that after a few generations, the political landscape as represented by the world map will no longer resemble historical reality (and in the very unlikely event that it does, this would have to be rather ascribed to serendipity than to intention).

A recent textbook defines history as "the study of people and the choices they made" (Schrag 2021: 18–21). While here is not the place to assess the validity of such simplifying assumptions, the definition seems nonetheless fitting for the experience of history in CK3, where most in-game events depend on individual choices. Historical research, however, is obligated to retain a sense of objectivity, even if this ideal can never be fully attained. And it is exactly this imperfection that enables historians to formulate objectivity as an *epistemological* problem in the first place (Ricœur 1988: 264). Instead of reconstructing medieval history in its entirety, the goal instead lies in the ongoing constructing of "contemporary images of the Middle Ages" (Goetz 2003: 265) based on sources and established methods. This is why alternative history remains a contested concept, even though it is widely accepted that contingency plays a significant role in historical storytelling, as Reinhart Koselleck (2015 [1979]) so prominently emphasised. Historical knowledge exists within the realm of human freedom, implying that it is always possible to speculate that things could have turned out differently. Acknowledging this fundamental contingency at the heart of all historical facts can serve as a methodological tool for identifying which factors were more decisive for any given historic event than others, by weighing all possible alternatives against each other: "The historian [...] must always maintain towards his subject an indeterminist point of view. He must constantly put himself at a point in the past in which the known factors still seem to permit different outcomes. If he speaks of Salamis, then it must be as if the Persians might still win." (Huizinga 1957: 292)

This, however, does not justify the use of counterfactual scenarios to reimagine events commonly regarded as 'decisive moments' as the branching points for the construction of alternative histories. At best, such scenarios may be justifiable as thought experiments restricted to narrowly defined time periods, as they are, generally, very difficult, if not impossible, to verify. The sheer number of potential alternatives is nearly unmanageable (Evans 2013; Peschke 2014: 229–238). By contrast, in modal logic and philosophy, counterfactual analysis is a well-developed and grounded method: "'Possible worlds' are total 'ways the world might have been', or states or histories of the *entire* world. [...] A practical description of the extent to which the 'counterfactual situation' differs in the relevant way from the actual facts is sufficient; the 'counterfactual situation' could be thought of as a miniworld or a ministate, restricted to features of the world relevant to the problem at hand." (Kripke 2001 [1972]: 18) However, as Samuel Kripke himself later admits, the terminology of 'possible worlds' and 'histories' can be misleading (ibid.: 18–20) and does not support a direct application to historiography.

4. Cultures in *Crusader Kings III*

The preceding reflections on counterfactuality, historical accuracy, and authenticity allow us to narrow the investigation to a more precise question: engaging with a past culture entails more than aggregating and analysing its material remnants. It means giving its spectres a place within the present. And to feel at home with these spectres, objective authenticity is not required – subjective authenticity will suffice. What players need is for their in-game experience of past cultures to feel 'recognisable' and convincing. The counterfactual possibilities offered by CK3 allow players to explore how the game system defines and constrains the meaning of 'culture'. On a basic level, this involves examining how game mechanics shape cultural expression. On a more reflexive level, the perceived limitations of these mechanics can serve as a model for understanding history itself: that it never actualises all that was possible but may always select a few outcomes from a broader field of potentiality. Playing with 'what if' scenarios of culture might then open a window onto what cultures *could have been*, while simultaneously confronting players with the necessities of why they developed the way they did.

When analysing CK3's complex culture system, however, it is important to recognise that 'culture', as we understand it today, is anachronistic when applied to the Middle Ages – at least as long as we focus on Europe. With its reference to manners, erudition, and learnedness, the classic Arabic term *adab*, for instance, did contain connotations that, in many ways, resemble what we today would understand as 'culture'.[2] In fact, an Almohad interpretation of *adab* might have influenced some of the core ideas on erudition as cultivated at the court of King Alfonso X of Castile and León (r. 1252–1284), who is worth mentioning here for having commissioned a *Book of Games* (*Libro acedrex dados e tablas*, c. 1284), one of the most important written sources on medieval game culture (cf. Fierro 2009; Conrad 2022: 49–72, esp. 72). That said, the term 'Middle Ages' can itself be regarded as Eurocentric in nature, since it conveys a notion of the period as bridging antiquity and early modernity, whereas in the Islamic World the period has been often referred to as the 'Golden Age' (s. Bauer 2020). Consequently, despite all its efforts to provide a more diversified view of the period, CK3 does not manage to fully erase all traits of its own cultural origins. This becomes even more apparent in that the specific form of its struggle with postcolonialism is the result of communicating from within a European, if not even Eurocentric, framework.[3] This paradox – that the criticism of Eurocentrism is formulated within a Eurocentric framework – indicates how difficult, if not even impossible, it is to leave behind cultural imprints. It therefore seems inevitable here to concentrate on European notions of culture and how they may have informed the game's representation and mechanics of culture.

Although the Latin term *cultura* was known to scholars, its scope was mostly confined to agriculture. Usual markers for identity constructions instead were, amongst others, religion, language, erudition, power, wealth, estate. These categories were far from homogenous; even religious practices could vary significantly from region to region, especially in areas where strong local traditions, often of pagan origin, persisted. Consequently, more recent approaches in cultural history have challenged the idea of a monolithic and static medieval Latin-Christian culture (Przybilski 2010: 10). The underlying notion of cultures as enclosed bodies was prominently developed by German scholar Johann Gottfried Herder (1744–1803). Herder conceived of cultures as monadic, homogenous spheres hermetically sealed off from the external world. In his view, they are defined by an orientation towards alterity, difference and exclusion. Encounters of two cultures would either lead to the assimilation of similarities or the repulsion of differences. According

---

2   I thank James Wilson (University of Konstanz) for having reminded me of this important issue.
3   Another aspect of the game that will not be further explored here but that similarly expresses this latent Eurocentrism is CK's fixation with Christian-feudal logics, for instance, with respect to the administrative organisation of political bodies even in non-Christian realms.

to Wolfgang Welsch (2017: 11), this mechanistic and tribalistic model of culture became extremely influential thereafter and could even be regarded a precursor to Samuel P. Huntington's 1993 slogan of a 'clash of civilisations'.

Therefore, it is little surprising that monolithic concepts of culture did influence computer games as well. *Civilization III* (Firaxis 2001), the third instalment of the famous game series invented by Sid Meier, is among the first historical games that introduced 'culture' as a game mechanic. In the game, 'culture' serves as a metric for the impact a civilisation has on people living in the environs of cities by means of customs, arts, and philosophy, that is, by advancements that are not part of the natural sciences or technology and have nothing to do with day-to-day survival. This 'culture' is generated by specialised buildings, such as temples, cathedrals, libraries, or universities. As such, the concept represents an accumulable resource, with the cultural value of a civilisation being represented by the sum of cultural points of all the cities that belong to this civilisation (which, upon closer inspection, seems a rather absurd idea: How could one possibly quantify the cultural significance of an entire civilisation?). Along with this concept of 'culture' also came a new victory type, 'cultural victory', which could be achieved under the following conditions: (1) one of the player's cities reaches 20,000 culture points, or (2) the civilisation surpasses a global cultural threshold (usually 100,000 points) and the overall sum is at least double that of any other civilisation.

This specific concept of culture was then continued in *Civilization V* (Firaxis 2010) and *Civilization VI* (Firaxis 2016), where social policies are unlocked with culture points that were now organised in their own 'technology trees'. The conceptualisation of culture as a quantifiable and assignable good exclusively belonging to a civilisation and not being shared with others reproduces monolithic notions linked to nationalism and neoliberal ideas related to the commodification and quantification of public goods. Which goes well with the game's title, given that the French word *civilisation* is an antonym to 'culture' whereas 'culture' is conceived as something holistic sensual, and organic, 'civilisation' refers to something abstract, alienated, mechanistic and utilitarian (Przybilski 2010: 11).

In contrast, the representation of culture in CK3 does not fit this mould. As Jason Pituzzello (2014) already noted for *Deus Vult* (Paradox 2007), an expansion to the game series' first edition, the game is renowned for its more elaborate ideas on culture:

"Culture, as defined in the context of *Crusader Kings*, is a label placed on a shared language and customs, excluding religion. Religion comprises a character of province's theological belief, excluding shared language and social customs. Character, but not provinces, also possess DNA, which is not a mapping of their genome so much as it is a sequence of numbers assigned to the various permutations of the game's character portraits. When children are born in the game, they borrow DNA from both their mother and father, producing character portraits that give members of the same family similar appearances. It also means that culture does not impact the shape of cheeks or noses. [...] Because technology can be transferred between provinces, this ensures that learning and technology can spread throughout Europe as it did historically and that contact between cultures occurs not only through warfare. This system of culture, religion, provinces, and characters creates multiple sites of conflict as well as interaction." (Jason Pituzzello 2014: 46)

While Pituzzello's description still mostly holds true for CK3, the culture system has, in the meantime, become even more complex and dynamic, which was also a reaction to that it used to feel "a bit static" (CK3 Dev Diary #64, n. p.) in comparison to other sub-systems, such as the faith system. Instead, the developers wanted 'culture' to correspond more to the overall design goal of offering players not only 'diversity' but also "a deeply immersive experience with more dynamic elements and player choice than ever before" (Dev Diary #22, n. p.).

In CK3, four key resources govern most in-game actions: gold, prestige, piety, and renown. Culture, by contrast, is not a resource, but a higher function with strategic impact on other game elements. Its primary function, however, is to determine which advancements – such as technologies and customs – become available to players. Additionally, culture can affect resource acquisition, research speed, and other variables. Each culture belongs to a *culture group*, that is, an aggregation of culturally proximate yet distinct cultures. This proximity influences the metric of 'cultural acceptance', which determines how characters from different cultures perceive and relate to each other.

Moreover, each culture in CK3 has one special character as its *culture head*, usually the character with the most counties of that culture within their sub-realm. Only this culture head has the power to reform his or her culture by adding new traditions (see further below), and to choose an innovation to be researched. Innovations represent technological, legal, and ideological advances, which are grouped into three categories: military and civic, cultural, and regional. Furthermore, each innovation belongs to one of four eras: tribal, early medieval, high medieval, or late medieval. Once at least half of an era's innovations are unlocked and the required minimum year is reached (476, 900, 1050, 1200, respectively), the culture will move on to the next era. For example, the high-medieval innovation 'Knighthood' will only be available as early as in 1050. Once achieved, it gives bonuses to armies, accolades and individual prowess. The innovation 'Banking', on the other hand, also available in the high-medieval period, will increase development growth in counties by ten percent. These examples illustrate the developers' effort to align innovations with the historical periods and regions in which they emerged. The regional innovation 'Sanitation', for instance, increases plague resistance and can only be discovered by Arabic or Byzantine cultures, thereby implicitly referencing the advanced medical standards in these societies at the time.

Innovations can become available through 'exposure' to neighbours who have already discovered them. As written in the developers' diaries, "(e)xposure is a natural process, occurring when your culture has counties that border another culture with a specific innovation. The more you have in common (culture group, religion, and so on) with that other culture, and the more of its counties your culture borders, the faster you'll unlock that innovation." (CK3 Dev Diary #27, n. p.) Describing this process as 'natural' highlights its random and involuntary character, mimicking the unintentional effects of intercultural contacts. Once an innovation is discovered, it can be used by any county and character of the culture that unlocked it. The learning skill of a culture head also influences how quickly an innovation will be reached, further integrating CK3's RPG approach into the simulation of dynamic cultural development.

Each culture rests on five cultural pillars: 'heritage', 'language', 'aesthetics', 'martial custom', and 'ethos'. 'Heritage' refers to the origins of a culture and is therefore mostly identical with its culture group. 'Language' determines the language spoken by its members and affects interactions between characters; if they share the same language they are more likely to develop positive relations. Also, the default birth names of characters, the names of titles and knights correspond with this language. 'Aesthetics' determine a culture's visual details, such as coats of arms, clothing styles, unit armour, architectural design, and available holding types. 'Martial customs' govern what genders can act as commanders or knights. 'Ethos' captures a culture's underlying values and worldview and, like its real-life counterpart, thus represents a culture's foundation. There are seven available *ethoi*: 'bellicose', 'bureaucratic', 'ceremonious', 'communal', 'egalitarian', 'spiritual', and 'stoic'. Each ethos provides gameplay bonuses and affects the ease of adopting specific traditions. All five pillars, along with traditions, influence a culture's acceptance value, from 0 percent (no acceptance) to 100 percent (full acceptance). Acceptance can also be influenced by modifiers and special events. In addition to these elements, each culture usually possesses four to six *traditions* (there is a fixed number of slots for each culture), chosen from a pool of 165, divided into five categories: realm, warfare social, ritual, and regional. Adopting a new tradition costs 2,000 prestige points, making cultural reform a rather late-game endeavour. In CK3, a culture thus emerges from a unique combination of all these components (heritage, ethos, traditions, language, martial customs, aesthetics, and traditions). Theoretically, the possible number of cultures is therefore vast. For example, multiplying *ethoi* and traditions alone (without considering other pillars) would yield 1,155 combinations. Yet not all are actually implemented: as of 1066, the analysis of game data (see further below) shows that there are 200 distinct cultures.

Given the large number of cultures in CK3, this section will focus on one single illustrative case: the 'Iberian' culture group at the 1066 start date, with particular attention to Catalan and Castilian (Fig. 1). This group is especially relevant due to its position on the contact zone between Christian and Islamic territories, making it a fruitful example for exploring in-game transculturality, cultural exchange, and hybridity. The focus will lie specifically on *ethos* and traditions, as these elements most strongly influence player strategies.

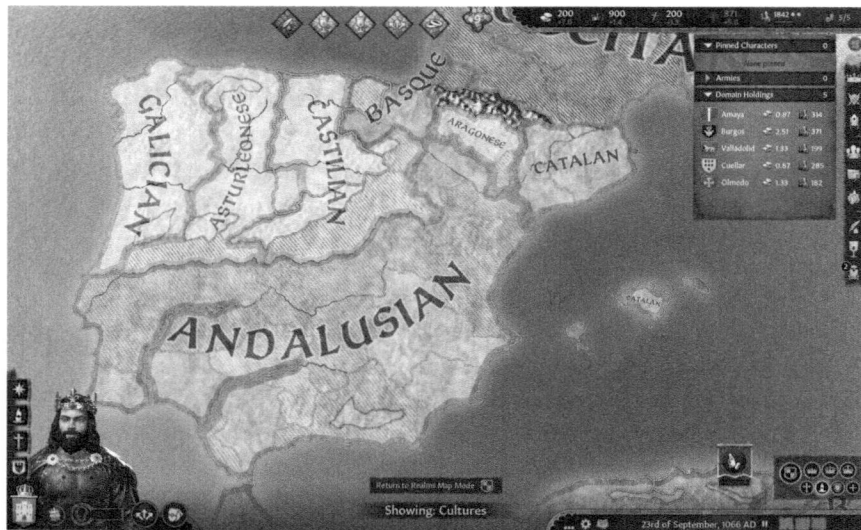

Fig. 1: *Cultures in Iberia for the start year of 1066. Transitional contact zones between cultures are represented as crosshatched areas.*

Catalan culture is characterised by a 'Ceremonious' ethos (Fig. 2), which the accompanying flavour text describes as follows: "The ceremony of court life is so integral to this culture that it forms a core part of people's social behaviour. A place for everyone, and everyone in their place." Mechanically, this ethos increases monthly prestige by 10 percent, reduces title creation costs by 15 percent, and grants a +5 opinion bonus among house members, while imposing a –10 opinion penalty on minority vassals. This reflects a focus on inner-court stability at the expense of cultural inclusivity. The choice of this ethos appears deliberate, very likely referencing King Peter IV of Aragon (r. 1319-1387), known as *El Cerimoniós*, 'the Ceremonious'. Peter IV is renowned for his patronage of the arts, the expansion of his realm overseas, the strengthening of crown authority and establishment of a strict court protocol.

Catalan culture has five traditions: 'Maritime Mercantilism', 'Parochialism', 'Refined Poetry', 'Visigothic Codes', and 'Ritualized Friendship'. 'Maritime Mercantilism' is a regional tradition only available to cultures located at coastlines encouraging sea trade and naval warfare. For it provides benefits such as +10 percent Coastal Holding Tax, –25 Sea Danger, a +10 percent increase of the opinion of Republic vassals, and earlier access to trade ports (Fig. 3). 'Maritime Mercantilism' is one of the few traditions whose introduction was announced in a special 'winter teaser' by the developers; according to them, one of the reasons for its inclusion was that it "makes coastal holdings more useful to a culture" (CK3 Winter Teaser #3, n. p.). Though the term 'Mercantilism' is anachronistic, the tradition reflects values of maritime commerce that fit to Catalunya and is well encapsulated by the motto: "The world may be ruled by armies, but this culture knows that it is truly controlled by whoever dominates the flow of gold across the seas."

Cultural Combinatorics and Conjured Spectres 97

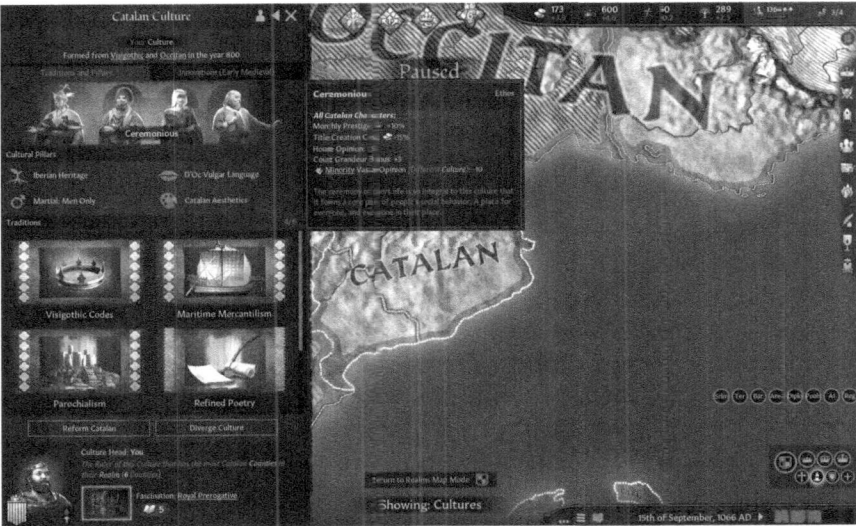

Fig. 2: The window shows all culture components of Catalan culture in 1066, with the pop-up window giving detailed information on the 'Ceremonious' ethos.

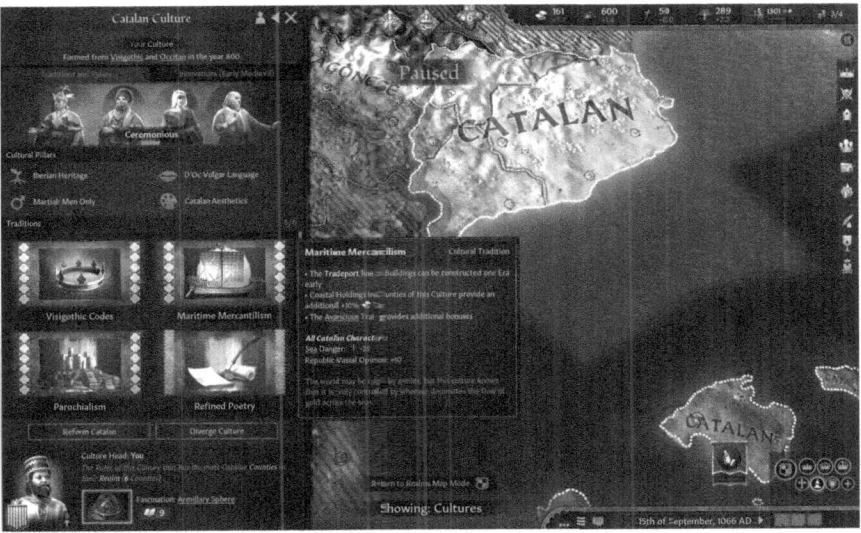

Fig. 3: The window shows all culture components of Catalan culture in 1066, with the pop-up window giving detailed information on the tradition of 'Maritime Mercantilism'.

'Parochialism', in turn, enhances the autonomy and competitiveness of cities: "City residents of this culture are fiercely competitive and independent. They invest a lot of money and energy to make sure that *their* city is the grandest." 'Refined Poetry' improves the court poet's abilities and the spread of legends, again reinforcing court-related issues. As the name suggests, the tradition of 'Visigothic Codes' is unique for Iberia and shared by Aragonese, Basque, Catalan, and Visigothic cultures. It grants immediate access to High Partition Succession Law and Equal Gender Law. The flavour text suggests that these cultures upheld an unbroken memory of Visigothic values and customs, even though this is not historically consistent. In CK3, Castilian, for example, lacks this trait, although it was especially Castile that claimed to be the true descendant of the Visigoths, while the presence of this tradition in the historically isolated Basque culture seems at least debatable. This is more the case since an informative text about the heritage of Catalan and Castilian explains that both "diverged from Visigothic in 800", alluding to the ruling power on the Peninsula prior to the Islamic conquest in 711. 'Ritualized Friendship', finally, facilitates building and maintaining friendships, which provide additional benefits. This tradition is shared by almost all cultures within what the game refers to as the 'Iberian' culture group (Andalusian, Aragonese, Asturleonese, Basque, Castilian, Galician, Portuguese, and Visigothic), except for the Suebi (who are, however, extinct at the game's earliest start date), indicating a shared cultural focus on personal alliances. That said, one must bear in mind that the in-game use of the label 'Iberian' is historically disputable, since it, with the exception of 'Andalusian', refers to Latin-Christian cultures as they had been present on the Peninsula throughout the Middle Ages. In a strict sense, however, Almohad, Almoravid, or the cultures of the small kingdoms after the collapse of the Caliphate of Córdoba after 1099 could similarly be addressed as *Iberian* cultures. According to the logic of CK3, cultures in the south of the Peninsula are either identified by the *passepartout* of 'Andalusian' (and therefore as part of the 'Iberian' culture group) or assigned to another culture group, such as Berber or Arabic. For reasons of clarity, in this article 'Iberian' is therefore set in quotation marks whenever the in-game category is referenced. All things considered, the inclusion of an 'Andalusian' culture – Arabic in origin, and with Islam as its religion – to the 'Iberian' culture group appears to reflect an effort to recognise the historical significance of Islamic cultures present on the Peninsula during the Middle Ages.

While shared traditions and membership in the same culture group increase cultural acceptance within the 'Iberian' group, the specific combination of cultural components also defines key aspects of cultural difference. For instance, Castile's ethos is 'Bellicose', promoting a distinctly martial orientation. It grants bonuses such as increasing prowess (+2) and levy size (+10 percent) as well as reduced costs for men-at-arms (-10 percent): "This culture considers conflict and violence to be necessary states of existence; ingrained in its people is the idea that one should stand up and fight for their own." Castile's traditions further reinforce this stance, with traditions like 'Castle Keepers', 'Chivalry', 'Martial Admiration', 'Hit-and-Run Tacticians', all encouraging expansionist land warfare. The 'Castle Keepers' tradition, in particular, aligns with the linguistic origin of 'Castile' (*castillo*, castle), and reflects its heraldic symbolism. This association has a deeper historic meaning, since Castile used to be a key military power in Iberia, by leading Christian campaigns against the Muslims territories, as exemplified by King Fernando III (r. 1217-1252), who was later canonised for his engagement in these military activities.

The brief comparison between Castilian and Catalan in CK3 illustrates how each culture's configuration of ethos and traditions acts as a strategic framework, guiding player behaviour by suggesting optimal paths while limiting others. As seen, Catalan culture incentivises courtly refinement, domestic government, and naval expansion, thereby mirroring historical developments such as the maritime empire of late medieval Catalunya/Aragon and the consolidation of crown authority. Castilian culture, by contrast, promotes land-based aggression with unmatched martial intensity; no other 'Iberian' culture contains as many military-oriented traditions. Each culture in CK3 thus acquires a distinct 'character' or 'identity' through its components, shaping playstyles by encouraging certain behaviours and discouraging others. While this does not prescribe player actions in a deterministic manner, it does create weighted incentives. Going 'against the grain' of a cultural identity is possible but often more difficult and less rewarding. Despite its simplification, this culture concept much resembles how cultural identities and norms subtly structure human agency: not through coercion, but through patterned tendencies and constraints.

5. Cultural hybridity in CK3

The discussion of culture in CK3 so far seemingly indicates that culture in the game would align with Herder's notion of cultures as monadic, self-contained entities. As cultural identities in CK3 are designed for recognisability and immersion, and though often grounded in historical evidence, this results in a tendency towards stereotypical representations. Prior to the release of the DLC "Royal Court" on 2 August 2022, this would have been a mostly accurate assessment, since cultures could only be altered through scripted decisions. What changed this situation radically was the introduction of cultural *hybridisation* and cultural *divergence*, which allowed players to adapt cultures more deliberately.

With these new rules, players can create a divergent culture at any point in the game. While a new ethos can be chosen freely, changes to languages, martial customs, and aesthetics, are more restricted. Aesthetics, for example, can only be modified if divergence has a historical precedent, such as in the case of Austria, which is a divergence from Central Germanic and comes with its own aesthetics for clothes, troops, architecture, etc. Divergent cultures benefit from a 50-year conversion bonus in counties of their parent culture, thereby facilitating cultural spread.

By contrast, hybridisation allows rulers to combine two cultures, provided the following conditions are met: (1) The other culture must be present within the realm, (2) both must have different heritages, (3) cultural acceptance must be at least 40 percent, and (4) both cultures must not have been created recently. Players must also select at least one pillar from each culture (Figs. 4 and 5). As explained in the developer's diary with the apt title "One Culture Is Not Enough" (22 June 2021), the goal is to combine cultures that are "different enough to warrant them being combined into a single culture, rather than just assimilating one in favour of the other." (CK3 Dev Diary #65, n. p.) As with divergence, the mechanics rely on principles of proximity, exposure, and acceptance to avoid arbitrary outcomes.

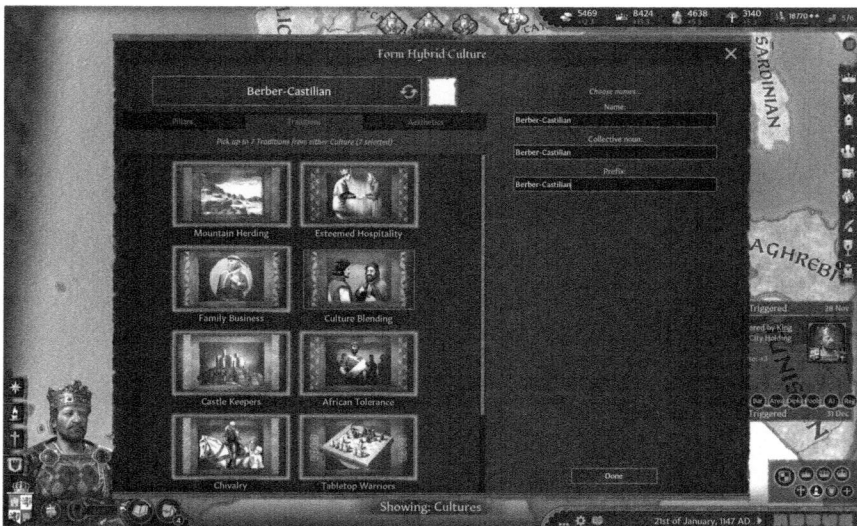

Fig. 4: The hybridisation of Castilian and Berber to a new Berber-Castilian culture in the year 1147. For that purpose, the player can here select and combine traditions from both cultures. Colours and names can be customised as well.

The main advantage of hybridisation lies in the ability to select heritage, cultural pillars, traditions, and aesthetics from both parent cultures. Hybrid cultures inherit all innovations known to their parents, including regional and era-locked ones. They also start with 100 percent cultural acceptance toward both parent cultures, which helps reduce internal tensions. By default, the names of hybrids are usually hyphenated (e.g., Siculo-Norman), although some hybrids have default names based on historical models. Hybrids-of-hybrids cannot be created, with the exception of special historical hybrids, such as English (a hybrid of Anglo-Saxon and Norman). Hybridisation is a very attractive strategy, especially if a ruler governs large territories made up of different cultures and wants to better integrate the conquered population to reduce the risk of uprisings. However, divergence and hybridisation require significant prestige (typically at least about 800 points) and take several in-game years to complete. These built-in barriers ensure that divergence and hybridisation do not happen too frequently nor too early in the game.

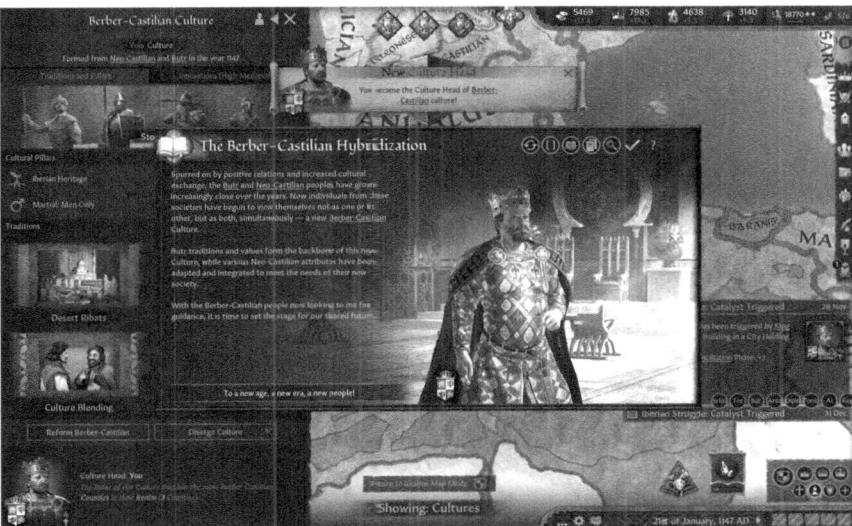

Fig. 5: After the successful creation of the new hybrid culture, the player is rewarded with this window. The text highlights cultural proximity, in-betweenness, and shared identity.

The integration of hybridisation into CK3 is remarkable, as the acknowledgment of the hybrid realities of medieval identities, particularly at transcultural "contact zones" (Burke 2009: 72), such as Iberia, Norman Sicily, the British Isles, or the broader Mediterranean (Jörg/Parker/Pleuger/Zwanzig 2011) is not common in historical games. The explicit use of the terms 'hybrid' and 'hybridisation' is revealing, as these are central to postcolonial discourse, which challenges Eurocentric, essentialist notions of culture and national identity in the light of globalisation.

Although the developers do not cite theorists such as Homi Bhabha, Edward Said, or Peter Burke directly, the conceptual vocabulary they use suggests familiarity with such frameworks. What is more, there are more straightforward indicators to be found in the developer diaries to other Paradox games. For example, the headline of a diary article on the game *Hearts of Iron 4* (Paradox 2016) discussing issues of decolonisation indeed uses the word "Post-Colonial" (HoI4 Dev Diary, n. p.), while a diary for *Victoria 3* (Paradox 2022), a game set in the Victorian Era, explicitly mentions "Orientalism" as an ideology of domination based on contrasting a rational West with a "stagnant" East, thereby echoing Said's theory of the same name (Victoria 3 Dev Diary #136, n. p.). This indicates an acquaintance of both developers and players with this discourse, which is further supported by the fact that Paradox encourages its community to provide them with input for identifying reappearing 'trends and issues' to improve their games (CK3 Dev

Diary #147), with some player comments demonstrating a knowledge of postcolonial and similar critical discourses.

Homi Bhabha famously introduced the idea of a 'third space', a liminal site of negotiation and ambiguity, where power relations can be subverted, and fixed meanings and identities can become fluid and contradictory (Bhabha 1995). Similarly, Edward Said promoted the notion of cultures as open, entangled, and heterogeneous entities (Said 2003 [1978]). Peter Burke's work, however, seems even more relevant to CK3, as he describes hybridity and transculturality as the result of sustained (inter–)cultural practices (Burke 2009; Welsch 2017: 12; Conrad 2022). In his seminal book on cultural hybridisation, Burke catalogues aspects and terminologies of hybridisation to support his thesis that cultural encounters spark innovation through processes of imitation and appropriation, accommodation and negotiation, cultural mixing, syncretism, cultural translation, and creolisation. At the same time, Burke acknowledges the ambivalence of hybridisation: while it fosters innovation, it may also provoke cultural loss and social conflict:

"The price of hybridization, especially the unusually rapid hybridization that is characteristic of our time, also includes the loss of regional traditions and of local roots. It is surely no accident that the present age of cultural globalization, sometimes viewed more superficially as 'Americanization', is also the age of reactive nationalisms or ethnicities." (Burke 2009: 7–8)

Nevertheless, Burke argues that such negative effects are often short-term and overshadowed by positive long-term outcomes. In turn, any attempt to create cultural purity can only be pursued with enormous efforts and is usually unsuccessful (Burke 2008: esp. 118–123, 130–142). Following Burke's lead, the ambivalence of hybridisation is reflected in CK3 as well, where cultural acceptance, proximity, and exposure can influence the success or failure of hybrid cultures (Burke 2009: 79–101).

6. A Brief data analysis of cultures in CK3

After these theoretical remarks, it seems appropriate to take an even closer look at some of the structural dispositions of the game system and how they not only shape the way in which players experience culture and intercultural contact, but also how they may inform strategic actions. That, exactly, is the objective of the following brief overview of some initial findings from an ongoing quantitative analysis of game data. This data is taken from a playthrough starting in the year 1066, with the player assuming the role of King Sancho II of Castile (r. 1065–1072). Data analysis and visualisation were performed on converted CK3 save files by using a self-written Python algorithm.

The first diagram (Fig. 6) shows the distribution of cultures at the initial state of the game, by counting the frequency of active, playable characters (including the player character), reduced to the first 50 cultures for reasons of clarity. According to the data analysis, there were altogether 22,437 living characters at this point, whose cultural identities, however, are not distributed equally: There is a clear peak of Greek culture with 916 characters, which is not coincidental, as this relates to the many characters living in the Byzantine Empire and its descendant realms, which together cover a great portion of the map and include many counties. After Greek, the next largest cultures are Ashkenazi and Bedouin, which might seem surprising at first. A possible explanation is that, in the game, Ashkenazi Jews are scattered all over the map due to their diaspora, while Bedouins live in many different regions within the vast Islamic world. The first-ranked European cultures are French (9th) and Franconian (10th), which belong to the – in terms of the number of counties – two largest (western) European realms, the Kingdom of France and the Holy Roman Empire. In contrast, Castilian, the culture of King Sancho II, comes in at place 48 with only 152 characters.

In the second diagram (Fig. 7), however, we find a different picture. Here, the cultures have been aggregated to their culture groups. Again, for reasons of clarity, only the top 30 culture groups are shown. The three culture groups with the most characters are 'Turkic' (1,752), 'Arabic' (1,517), and 'Indo-Aryan' (1,512), whereas the 'Iberian' culture group (to which Castilian belongs) holds the 7th position and therefore ranks much higher than in the previous diagram. Characters of the 'Turkic' ethnic group live across Central Asia, while 'Indo-Aryans' are concentrated in Northern India. 'Arabic' cultures are found all over the enormous Islamic world that covers great parts of Asia, Africa, and Iberia. In turn, the sum of characters of the 'Iberian' group reflects both the multi-cultural situation as well as the great number of rather small counties in the Peninsula. This further underscores the developers' effort to create cultural distributions that mirror historical realities, also by simulating existing cultural world regions such as the Islamic world.

All things considered, the data analysis confirms something that players typically experience more intuitively: some cultures and culture groups (such as Arabic, Greek or Iberian cultures) are more dominant by numbers than others, thereby offering strategic advantages, especially with respect to cultural change, and making them more attractive choices for players. Moreover, while larger culture groups tend to enjoy more cultural stability, they might incentivise players to create divergent cultures that make it easier for them to become a cultural head. In contrast, choosing a character in a border county exposes them to foreign cultures, which can promote progress if both cultures have a high acceptance rate but may have negative effects if they do not. Over time, this can make hybridisation a more appealing and more likely option.

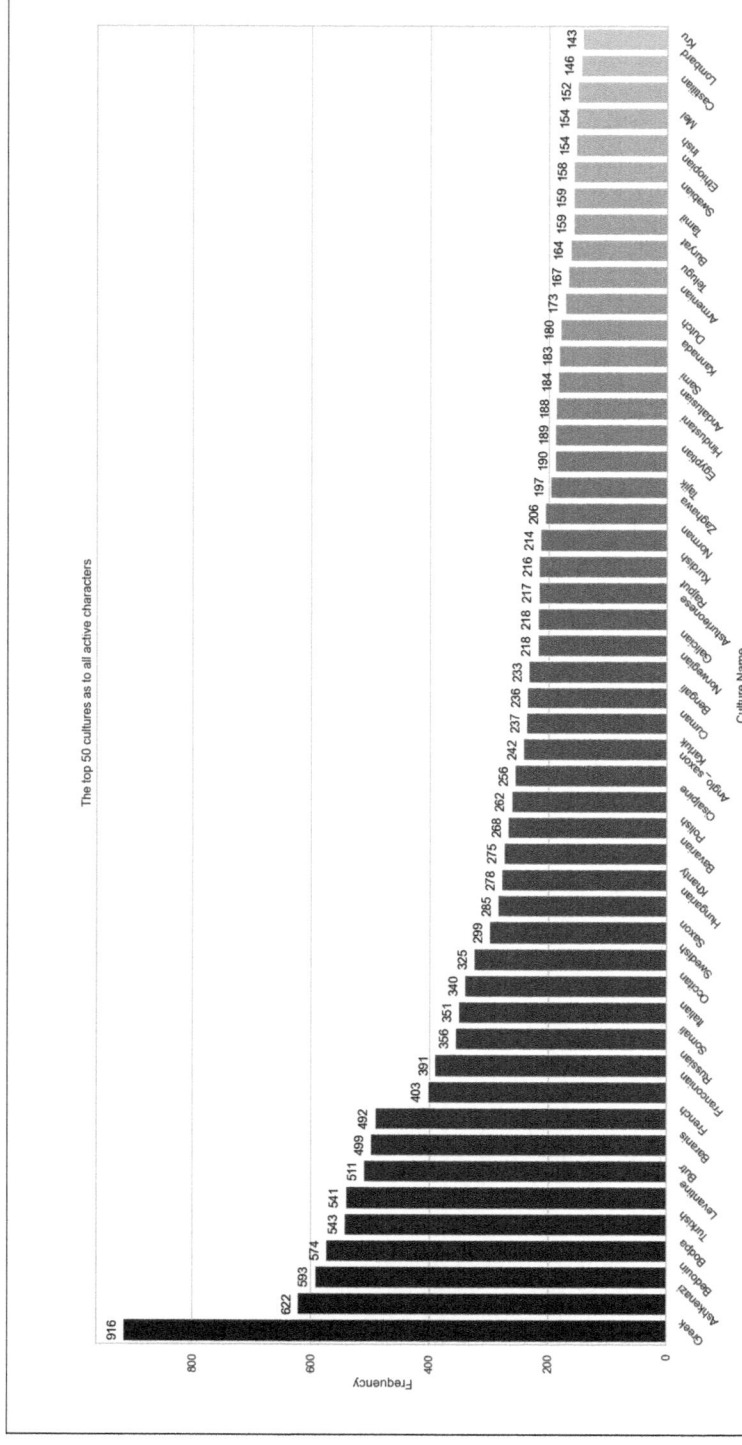

Fig. 6: The top 50 cultures according to the numbers of active characters in Crusader Kings III for the start year of 1066.

Cultural Combinatorics and Conjured Spectres

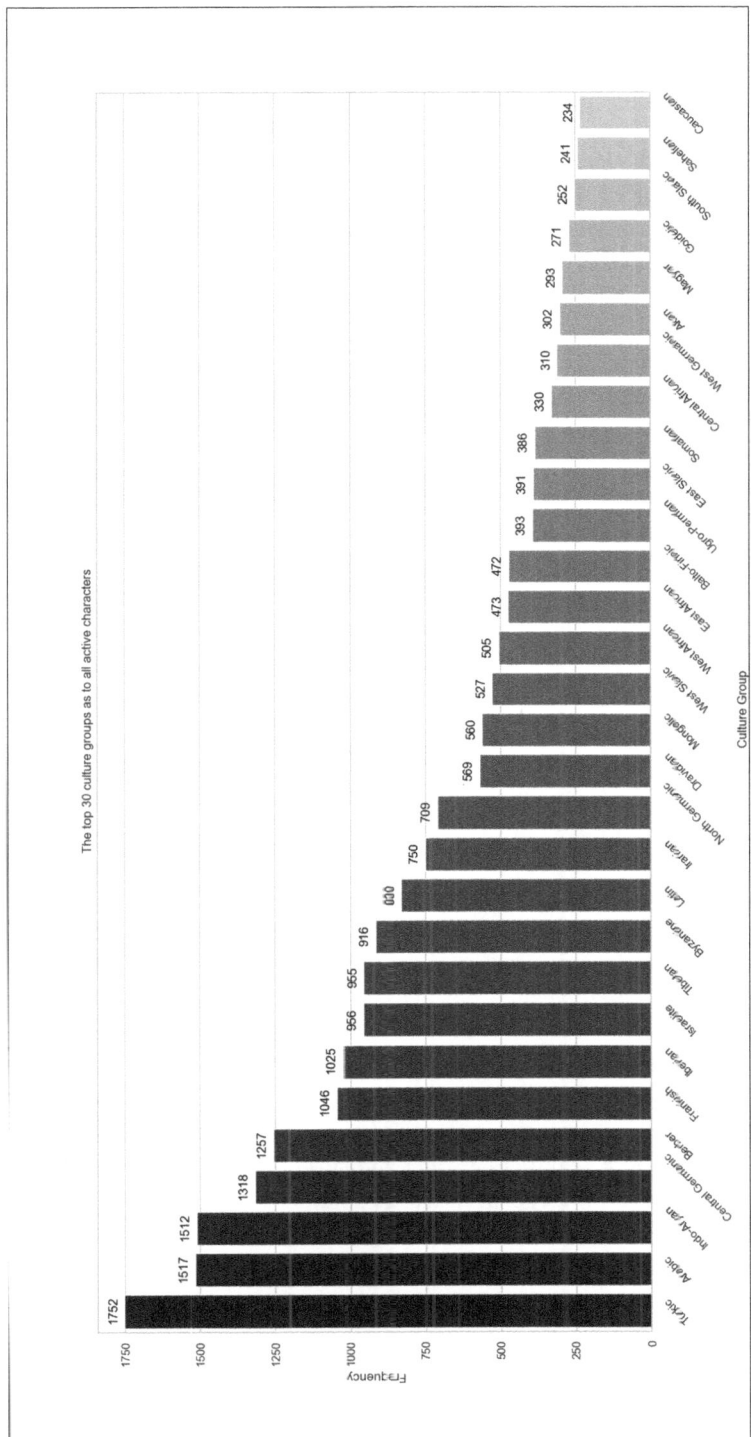

Fig. 7: The top 30 culture groups according to the numbers of active characters in Crusader Kings III for the start year of 1066.

7. Conclusion

As demonstrated in the preceding sections, the current version of CK3 places strong emphasis on cultural encounters and transformation. Unlike many other medievalist media, the game presents players with a period defined by intercultural relations that act as the driving force of change. The complexity of cultural dynamics has been a distinctive hallmark of the CK series since its inception, passed down through successive generations of designers and constituting a core element of its design philosophy (Pitruzzello 2014: 51).

CK3 treats culture not as a mere resource within the path dependencies of a technology tree but as a composite system embedded in a referential framework linking character traits, resources, eras, and intellectual principles. Conceptually, 'culture' thus operates as a transcendental 'condition of possibility', enabling higher-level functionalities and expanding strategic agency (Nguyen 2022: 78). This quality mirrors its real-world counterpart: likewise, culture facilitates innovation and transformation within a society. Yet while CK3 allows culture to behave as an open concept, this openness is not consistent across all levels. The game's spatial logic assigns a single culture to each county, fixed by map design (Fig. 1), with the only exception being transitional phases during shifts between two cultures (visualised on the map as crosshatched areas). In this regard, CK3 still remains locked in the framework of cultures as monolithic building blocks.

This tension is the result of a design trade-off. To increase a game's playability and attractivity for larger audiences, complexities of reality must be reduced, which is mostly done by translating them into the discrete and quantitative values a computer can handle more easily. One side effect of this quantification process is that it nudges players toward rational choice models that focus on optimising (quantified and quantifiable) utility, potentially favouring expansionist behaviour. This is one reason why many strategy games revolve around resource management or systems that resemble the – now mostly obsolete – models of econometry or 'cliometry', which once sought to ground historical research in mathematical theories and data analysis (Heinze 2012: 241, 302–303). However, this simplification does not diminish the game's artistic value. On the contrary, CK3 succeeds in crafting a rich and vivid historical experience within the constraints of the medium. Moreover, by appealing to broader audiences, it can support the democratisation of historical knowledge, provided that players are encouraged to reflect critically on fact versus fiction. Given that digital games are especially popular among younger generations, the game could be used as a starting point for classroom discussions.

At the same time, the strategic instrumentalisation of culture represents it as something that can be formed at will or even weaponised. As much as this fits the god-like perspective of a strategy game, it still deviates from historic reality. The hybridisation mechanic incentivises players to combine cultures for faster access to innovations or for harmonising inter-cultural tensions in large, multi-cultural realms. While there are historically grounded examples of cultural hybridity, most famously in Sicily with its unique, fourfold, Norman-Arab-Byzantine-Latin architecture as exemplified by the *Cappella Palatina* in Palermo (Conrad 2021; Conrad 2022), CK3 presents hybridity as an intentional and strategic choice, which is a modern notion far removed from the often slow and mostly involuntary cultural shifts of the medieval world. The game tries to balance this *mechanistic* view, quite ironically, with *mechanics* like peasant acceptance, so that a culture will only spread across a realm if it is accepted by the ruler's subjects, who can try to force same ruler to convert to their culture through rebellions. However, as user vilkas622 commented on June 22, 2021, "[I]t still seems to me to be too easy to create a hybrid culture. Most of the time when kingdoms acquired territory in this age, the ruling elite may change to the new kingdom and their culture, but actually changing the culture of the peasants was not something that happened." (CK3 Dev Diary #65, n. p.) As they point out, hybridity can be applied rather deliberately, which makes it appear more frequent than it historically was.

Nevertheless, the engagement with postcolonial concepts is noteworthy. Terms like hybridisation, cultural acceptance, and proximity introduce fluid cultural identities that may challenge Eurocentric essentialism. Drawing on Edward Said, Souvik Mukherjee has argued for counterstrategies in gameplay, such as direct protest, counterplay, and critical reinterpretation, to confront colonial ideologies. Yet any form of resistance within a game remains constrained by the medium itself, by its structure: "Alternative narratives can be written into being in the game world but only within the system that the game provides." (Mukherjee 2016: 518) This applies to CK's culture system, where alternative identities remain tied to the logic of the game system. Still, by expanding the scope of player autonomy and agency (Nguyen 2022), CK3 marks a significant shift in how culture is represented in historical games.

At the same time, however, CK3 largely omits key historical practices for imposing cultural purity, such as active discrimination or expulsion, as was the fate of the Castilian Jews after the Alhambra Decree of 1492. In fact, CK2 (Paradox 2012) had once included the expulsion of Jews as a game mechanic, which is not to suggest moral approval thereof, but rather to point out a tension: despite the series' general commitment to historical accuracy, this particular exclusion seems to prioritise *contemporary* moral and political sensitivities over historical reality. In this way, CK3 may at times reveal more about our present days than our past, which is also the result of the paradox that the implicit criticism of Eurocentrism is inevitably formulated within a Eurocentric framework. Overall, this highlights the importance of the central feedback loop between historical knowledge and its ludic representation. Game mechanics can subtly encode values and shape our understanding of history by organising, incentivising, and directing player actions – often without their conscious awareness. By inviting players to engage with cultural dynamics through play, CK3 offers a powerful model of historical imagination. In doing so, it not only enables us to engage with the spectres of the past, but perhaps even develop our very own 'politics of memory, of inheritance, and of generations'.

References

Addriaansen, R. (2022): "Conceptualizing the Period Rush: Dimensions of Immersive History in Historical Reenactment." In: M. Carretero/B. Wagoner/E. Perez-Manjarrez (eds.), Historical Reenactment: New Ways of Experiencing History. New York and Oxford: Berghahn, pp. 33–38.
Aust, H. (2009): Der Historische Roman. Stuttgart: Metzler.
Bauer, T. (2020): Warum es kein islamisches Mittelalter gab. Das Erbe der Antike und der Orient. Munich: C. H. Beck.
Brown, D. W. (2012): "Game-playing-role. The Suspension of Disbelief in Videogames." PhD thesis, School of Arts, Brunel University Press.
Bhabha, H. K. (1995): "Cultural Diversity and Cultural Difference." In: B. Ashcroft/G. Griffiths/H. Tiffin (eds.), The Post-Colonial Studies Reader. London: Routledge, pp. 206–209.
Burke, P. (2009): Cultural Hybridity. Cambridge and Malden: Polity.
Burke, P. (2008): What is Cultural History? Cambridge and Malden: Polity.
Buse, P./Scott, A. (1999): "Introduction: A Future for Haunting." In: P. Buse/A. Scott (eds.), Ghosts: Deconstruction, Psychoanalysis, History. London: Palgrave, pp. 1–20.
Caillois, R. (2001 [1958]): Man, Play and Games. Urbana and Chicago: University of Illinois Press.
Carretero, M./Wagoner, B./Perez-Manjarrez, E. (2022): "Introduction. Approaching Historical Reenactment." In: M. Carretero/B. Wagoner/E. Perez-Manjarrez (eds.), Historical Reenactment. New Ways of Experiencing History. New York and Oxford: Berghahn, pp. 1–18.
"CK3 Dev Diary #0: The Vision" (29 January 2025). Retrieved from https://forum.paradoxplaza.com/forum/developer-diary/ck3-dev-diary-0-the-vision.1265472.
"CK3 Console Dev Diary #2: Discovery in the Crusader Kings Experience" (29 January 2025). Retrieved from https://forum.paradoxplaza.com/forum/developer-diary/ck-console-dd-2-discovery-in-the-crusader-kings-experience.1515647.
"CK3 Dev Diary #22: A Medieval Tapestry" (14 April 2020). Retrieved from https://forum.paradoxplaza.com/forum/developer-diary/ck3-dev-diary-22-a-medieval-tapestry.1377711.
"CK3 Dev Diary #27: Cultures & Cultural Innovations" (31 January 2025). Retrieved from https://forum.paradoxplaza.com/forum/developer-diary/ckiii-dev-diary-27-cultures-cultural-innovations.1391886.

"CK3 Dev Diary #28: Art Focus" (May 26 2020). Retrieved from https://forum.paradoxplaza.com/forum/developer-diary/ckiii-dev-diary-28-art-focus.1395627.

"CK3 Dev Diary #64: Cultures Are Forever" (June 15 2021). Retrieved from https://forum.paradoxplaza.com/forum/threads/ck3-dev-diary-64-cultures-are-forever.1479565/.

"CK3 Dev Diary #65: One Culture Is Not Enough" (31 January 2025). Retrieved from https://forum.paradoxplaza.com/forum/developer-diary/ck3-dev-diary-65-one-culture-is-not-enough.1480529.

"CK3 Dev Diary #147: Some Words from Our Game Director" (30 April 2024). Retrieved from https://forum.paradoxplaza.com/forum/developer-diary/dev-diary-147-some-words-from-our-game-director.1673686/).

"CK3 Community Wiki." Retrieved from https://ck3.paradoxwikis.com/Crusader_Kings_III_Wiki.

"CK3 Winter Teaser #3: Cultural Traditions" (December 21 2021). Retrieved from https://forum.paradoxplaza.com/forum/developer-diary/winter-teaser-3-cultural-traditions.1503317.

Conrad, M. A. (2021): "Vertical Diffusion and the Patrons of Urban Palaces in 14th-Century Toledo." In: Giese, F. (ed.), Mudejarismo and Moorish Revival in Europe. Cultural Negotiations and Artistic Translations in the Middle Ages and 19th-Century Historicism. Leiden and Boston: Brill, pp. 108–132.

Conrad, M. A. (2022): Ludische Praxis und Kontingenzbewältigung im Spielebuch Alfons' X. und anderen Quellen des 13. Jahrhunderts. Berlin and Boston: De Gruyter.

Conrad, M. A. (2022): "Tropes of Transculturality: On Comparing Medieval Sicily and Castile." In: M. Ö Altınöz (ed.), Cultural Encounters and Tolerance Through Analyses of Social and Artistic Evidences: From History to the Present. Hershey: IGI Global, pp. 1–19.

Derrida, J. (1994): Specters of Marx The State of the Debt, the Work of Mourning, and the New International. New York and London: Routledge.

Evans, R. J. (2013): "Possible Worlds." In: Altered Pasts: Counterfactuals in History. London: Brandeis University Press, pp. 92–126.

Fierro, M. (2009): "Alfonso X 'the Wise': The Last Almohad Caliph?" Medieval Encounters 15, pp. 175–198.

Firaxis (2001): Civilization III [Video game]. Firaxis.

Firaxis (2010): Civilization V [Video game]. Firaxis.

Firaxis (2016): Civilization VI [Video game]. Firaxis.

Fisher, M. (2013): "The Metaphysics of Crackle: Afrofuturism and Hauntology." Dancecult 5(2), pp. 42–55.

Frasca, G. (2003): "Simulation versus Narrative: Introduction to Ludology." In: M. Wolf/B. Perron (eds.), The Video Game Theory Reader. New York: Routledge, pp. 221–235.

Heinze, C. (2012): Mittelalter Computer Spiele: Zur Darstellung und Modellierung von Geschichte im populären Computerspiel. Bielefeld: transcript.

Goetz, H.-W. (2003): "Geschichtswissenschaft und Geschichtsbewusstsein: Gegenwärtige Tendenzen der Mediävistik." In: R. Ballof (ed.), Geschichte des Mittelalters für unsere Zeit. Stuttgart: Steiner, pp. 265–278.

"HoI4 Dev Diary: A Post-Colonial World: Map Changes and New Tags" (11 July 2018), https://forum.paradoxplaza.com/forum/developer-diary/hoi4-dev-diary-a-post-colonial-world-map-changes-and-new-tags.1109323/.

Huizinga, J. (1957): "The Idea of History." In: F. Stern (ed.), The Varieties of History: From Voltaire to the Present. New York: Meridian.

Huizinga, J. (2002 [1938]): Homo Ludens: A Study of the Play-Element in Culture. London: Routledge.

Huizinga, J. (2014): "The Task of Cultural History." In: Men and Ideas, History, the Middle Ages, the Renaissance. Princeton: Princeton Press, pp. 17–76.

Jörg, C. /Parker, K. S./Pleuger, N./Zwanzig, C. (2011): "Soziale Konstruktion von Identität. Prozesse christlicher Selbstvergewisserung im Kontakt mit anderen Religionen." In: M. Borgolte/J. Dücker/M. Müllerburg/B. Schneidmüller (eds.), Integration und Desintegration der Kulturen im europäischen Mittelalter. Berlin: Akademie, pp. 17–102.

Juul, J. (2005): Half-Real: Video Games Between Real Rules and Fictional Worlds. Cambridge: MIT Press.

Koselleck R. (2015 [1979]): "Der Zufall als Motivationsrest in der Geschichtsschreibung." In: Vergangene Zukunft. Zur Semantik geschichtlicher Zeiten. Frankfurt am Main: Suhrkamp, pp. 158–175.

Kripke, S. A. (2001 [1972]): Naming and Necessity. Cambridge: Harvard University Press.

Leerssen, J. (2007): "Identity/Alterity/Hybridity." In: M. Beller/J. Leerssen (eds.), Imagology: The Cultural Construction and Literary Representation of National Characters. Amsterdam and New York: Rodopi, pp. 335–342.

Mukherjee, S. (2016): "Playing Subaltern: Video Games and Postcolonialism." Games and Culture 13(5), pp. 504–520.

Nguyen, C. T. (2022): "Games and Autonomy." In: Games: Agency as Art. Oxford: Oxford University Press, pp. 74–98.

Paradox (2004): Crusader Kings [Video game]. Paradox.

Paradox (2007): Crusader Kings: Deus Vult [Video game]. Paradox.

Paradox (2012). Crusader Kings II [Video game]. Paradox.

Paradox (2016): Hearts of Iron 4 [Video game]. Paradox.

Paradox (2020): Crusader Kings III [Video game]. Paradox.

Paradox (2022): Victoria 3 [Video game]. Paradox.

Peschke, H.-P. von (2014): Was wäre wenn. Alternative Geschichte. Darmstadt: Theiss.

Pitruzzello, J. (2014): "Systematizing Culture in Medievalism. Geography, Dynasty, Culture, and Imperialism in *Crusader Kings: Deus Vult*." In: D. T. Kline (ed.), Digital Gaming Re-imagines the Middle Ages. London and New York: Routledge, pp. 43–52.

Przybilski, M. (2010): Kulturtransfer zwischen Juden und Christen in der deutschen Literatur des Mittelalters. Berlin and New York: De Gruyter.

Ricœur, P. (1988): Zeit und Erzählung, vol. 1: Zeit und historische Erzählung. Munich: Wilhelm Fink.

Said, E. (2003 [1978]): Orientalism. New York: Vintage.

Schrag, Z. M. (2021): The Princeton Guide to Historical Research. Princeton: Princeton Press.

"Victoria 3 Dev Diary #136: Pivot of Empire Narrative Content" (7 November 2024), https://forum.paradoxplaza.com/forum/developer-diary/victoria-3-dev-diary-136-pivot-of-empire-narrative-content.1713939/.

Welsch, W. (2017): Transkulturalität. Realität—Geschichte—Aufgabe. Vienna: New Academic Press.

Zimmermann, F. (2020): "Introduction: Approaching the Authenticities of Late Modernity." In: M. Lorber/F. Zimmermann (eds.), History in Games. Contingencies of an Authentic Past. Bielefeld: transcript, pp. 9–24.

Playing Adewale
The Politics of History in *Assassin's Creeds Freedom Cry*

Osvaldo Cleger

Abstract

Released on December 17, 2013, Assassin's Creed: Freedom Cry is part of Ubisoft's Assassin's Creed series, designed as a historical simulation that explores the complexities of the 18th-century Caribbean slave trade. Departing from the franchise's typical monumental approach to historical content, catering to a 'touristic gaze', the game prioritises a politically charged narrative centred on Adewale, a former slave turned Assassin and Maroon leader. Adewale's quests during the game focus on liberating enslaved individuals and championing the Maroon movement within the French colony of Saint-Domingue, addressing themes of oppression, resistance, and solidarity that resonate with contemporary audiences.

Upon its release Freedom Cry was lauded by players and critics as a prime example of how video game stories can critically engage with historical and cultural narratives beyond the colonialist gaze that permeates much of the pseudo-historical content generated by the industry. The game's design, featuring missions like plantation liberations and the boarding of slave ships to free captives, creates immersive experiences that reconstruct the realities of slavery while empowering players to respond empathically by freeing the enslaved and confronting their oppressors. Player reactions, documented in streams and forums, demonstrate how the game sparked meaningful engagement with its themes and missions, encouraging real-world reflections inspired by the gameplay experience.

This article examines Assassin's Creed: Freedom Cry through the lens of what I term 'the politics of history', focusing on how its portrayal of the Haitian Revolution allegorically links historical struggles against slavery to contemporary movements for Black civil rights. By integrating analyses of history, social identity, politics, and empathic game design, this study evaluates the game's engagement with historical contexts, its ability to deliver meaningful gameplay, and its impact on culturally marginalised gaming communities.

Keywords

Historical simulation, video games on slavery, race relations in media, empathic game design, the politics of history

1. Introduction

Released on December 17, 2013, *Assassin's Creed: Freedom Cry* (Ubisoft 2013; hereafter ACFC) is a video game that simultaneously builds upon the industry's ongoing efforts to develop increasingly complex approaches to challenging historical subjects – such as the representation of slavery as a playable topic – while also departing from what has typically been offered by the video game industry in terms of historical memorialisation. ACFC presents a game narrative that exhibits both liminal continuities and abrupt departures within the tradition it occupies.

In this article, I engage both the game, situated within the broader landscape of the video game industry's portrayals of slavery, and recent commentary on ACFC's treatment of the topic, including Sarah Lauro's (2020) reflections on slavery in interactive media and Alyssa Sepinwall's (2021) chapter on *Freedom Cry* within her broader study of slavery and revolt in visual culture. While these contributions offer valuable context, my analysis develops a distinct line of inquiry focused on the game's formal, procedural, and spatial design. I argue that *Freedom Cry* departs from Ubisoft's customary monumental aesthetic – characterised by its emphasis on architectural grandeur and curated historical spectacle – and instead enacts what I here term a politics of history[1]. This approach

---

1   There is a longstanding tradition in historical inquiry that foregrounds the political forces shaping how history is written, remembered, and erased. From Ernest Renan's well-known assertion that nations are formed as much through forgetting as through remembering, to more recent historiographical interventions that examine the political construction of collective memory and the dynamics of forgetting and memorialising (Trouillot 1995; Wang 2012; Szaban et al. 2025), scholars have underscored the contingent and contested nature of the past. As Zheng Wang writes, "In the workings of social life, the past does not always exist as a hard, objective, or factual reality – something 'out there' to be grasped and appropriated. The past is not solid, immutable, or even measurable; rather, it is a fluid set of ideas, able to be shaped by time, emotion, and the politically savvy." (Wang 2012: 13) Building on this tradition, my use of the term *politics of history* departs from a focus on official historiographical practice and instead examines how contemporary cultural products – such as commercial video games – mobilise historical material not as inert content, but as part of an emotionally and politically saturated terrain. In this framework, *ACFC* can be read as a site where the legacy of slavery and colonial violence is reactivated within a mass-mediated format. Players do not approach this material as blank slates: their responses are shaped not only by the game's aesthetic strategies but also by the pre-existing popular narratives about history that they bring with them, narratives that reflect ideological bias and receive unequal institutional or cultural support. The decision to bring this game into conversation with the rise of the Black Lives Matter movement is not incidental: both phenomena reveal how historical memory is not only preserved or distorted through formal channels, but

foregrounds not the static reproduction of historical detail but the activation of political resonance, inviting players to engage not as tourists of the past but as participants in the unfinished struggles it evokes. Rather than constructing Saint-Domingue as a monumentalised space, ACFC, as we will see, strips its world of ornamental excess to centre the materiality of oppression and the immediacy of resistance. The game's reliance on empathic gameplay, its insistence on player identification with a formerly enslaved protagonist and with acts of collective liberation, functions not simply as narrative affect but as a formal intervention into the conventions of historical simulation. Through these mechanisms, the game unsettles the racialised logic often embedded in mainstream historical games and reorients player agency toward solidarity rather than domination. In support of this reading, I also draw on responses from players, game reviewers and cultural commentators, whose reactions to gameplay and reflections on the experience reveal how language, setting, and mechanics come together to produce moments of recognition and ethical engagement. This dual emphasis – on close analysis of the game's formal design and on a considered selection of responses from players and cultural commentators – structures the method I employ throughout the article. While not exhaustive, the sample reflects a sustained effort to survey and incorporate a wide range of available reactions. Its scope is shaped primarily by the limits of the article format – as opposed to a long-form monograph – and by the fact that many available gameplay videos and streams offer little or no verbal commentary from players.

As previously noted, ACFC both extends and disrupts earlier video game engagements with slavery. In terms of continuities, as Sarah Lauro notes in *Kill the Overseer!* (2020), the central theme of ACFC is far from new. Ubisoft had previously explored similar territory with *Assassin's Creed III: Liberation* (Ubisoft 2012, hereafter *ACL*), set in 18th-century French Louisiana, where the playable character Aveline de Granpré, the biracial daughter of an African slave, secretly works to liberate enslaved people. Beyond the Ubisoft franchise, this theme can be traced back to earlier video games from the 1980s, particularly in *Freedom: Rebels in the Darkness* (Coktel Vision 1988). In this 8-bit game, players assume one of several roles, including that of the famous Maroon leader François Makandal, offering one of the first playable representations of resistance against slavery in video game history.

In terms of departures, ACFC is the first game in the series to directly address the world of colonial slavery, making it the central theme of the simulation, the

---

also reconfigured and contested in popular culture, where affect, identification, and political urgency converge. Together, they exemplify the ways in which the politics of history unfolds beyond academic or state-sponsored discourse, extending into the digital and cultural spheres that increasingly shape public understandings of the past.

main driver of in-game quests, the most defining trait of the main playable character, Maroon leader Adewale, and a central element in experiencing most game mechanics. When compared to a game like *Assassin's Creed IV: Black Flag* (Ubisoft 2013; hereafter *ACBF*), set mostly in the colonial plantation economy of Cuba, the difference in how both games address the issue of slavery could not be starker. In ACBF, slavery is merely a backdrop for gameplay primarily focused on naval conflict among Caribbean pirates or stealth and combat missions undertaken by the main character, Welsh-born assassin and privateer Edward James Kenway, as he battles the Templars in Cuba and the surrounding Caribbean islands. By contrast, ACFC makes the quest of liberating slaves and fighting for their freedom the central mission players must complete.

In this regard, ACFC not only goes further than ACBF in its treatment of this subject – despite being a subsidiary of the latter – but also offers a depiction of colonial slavery and the Maroon movement of the 18th-century that is far less compromised than that of ACL. As Lauro points out, ACL displays "an absence of a radical abolitionist message for most of the game" (Lauro 2020: 39), and its main character finds herself often conflicted while playing the different roles assigned to her, which are signalled in the game by the clothes she wears, either as a lady from a wealthy family, an assassin fighting the Templars, or in her slave's rags.

ACFC is not only a unique instalment in the Ubisoft franchise and the gaming industry but also stands out as a significant media offering when viewed against the historical backdrop of representations of the Haitian Revolution and Haiti itself. As Raphael Hoermann has shown, portrayals of the Haitian Revolution and Black Haitians have long been shaped by what he terms the "Transatlantic Haitian Gothic," a racist colonial narrative that depicted Black Haitians as inherently evil: "a cross between savages, cannibals, predators, and Gothic vampiric 'monsters, thirsting after blood, and unsated with carnage'" (Hoermann 2016: 185). This narrative spread rapidly as news of the Black slave revolts and the overthrow of white masters in Haiti reached nearby colonies and European cities in the late 18th century. Unfortunately, this Gothic portrayal has maintained a troubling grip on the Western imagination, as evidenced by the ongoing demonization of Haitians in contemporary media[2].

In its representation of characters of African descent, ACFC is equally notable. Several scholars have highlighted how the video game industry remains dominated by White Euro-American culture (Williams et al. 2009; Brock 2011; Cleger 2016; Feagin 2020), reflected in assumptions about a predominantly white

---

2   A recent and particularly ugly example of this was unfolding as I was writing this article, with the then Republican candidate for the U.S. presidency and his vice-presidential running mate launching a campaign in Springfield, Ohio, to dehumanise Haitian immigrants, falsely accusing them of actions as grotesque as eating the pets of 'rightful' citizens and residents (see Thomas/Wendling 2024).

male audience, the dominance of white protagonists, a preference for white-centric narratives, and the frequent caricaturising or tokenisation of non-white characters. Against this backdrop, ACFC stands out by creating a predominantly Black open world where the primary mission centres on the well-being of non-playable characters (NPCs) of African descent and by featuring a Black hero who fully embodies this mission.

The context for the release of ACFC in 2013 is also highly significant. The game came out as the pro-Black civil rights movement in America gained momentum, marked by the founding of Black Lives Matter (BLM) that same year (Taylor 2016; Hillstrom 2018). Two years later, the #OscarSoWhite hashtag emerged, criticising the lack of diversity in the 2015 Oscar nominations and sparking broader scrutiny of an entertainment industry slow to change (Yuen 2016). It would take another three years for *Black Panther* (2018) to begin to address these demands for broader representation. While ACFC may not have matched the cultural impact of Marvel's *Black Panther* (Goren/Carnes 2023), Ubisoft's earlier effort to address representational inequalities deserves recognition.

The battle for representation in political and cultural arenas has defined the last decade. While movements like #OscarSoWhite and BLM operate in different spheres, both address issues of representation in contemporary society[3]. BLM, in particular, highlights the urgent need to improve societal perceptions of Black lives, advocating for greater respect, civil protections – especially from racist vigilantes and law enforcement – and policies that uplift this demographic. It also calls for a media culture that celebrates, rather than stigmatises, the achievements of Black citizens[4].

ACFC, a game whose main quest involves liberating African slaves, igniting a rebellion against white colonists in Saint-Domingue, and setting the stage for the Haitian Revolution, engages with a symbolic and cultural agenda that resonates

---

3   In a *Rolling Stone*'s article published by Touré, Patrisse Cullors, one of the movement founders, stated in this regard: "Many of us were tired and disturbed by the lack of recognition towards the killings of black people by vigilantes and law enforcement, [...] we were tired of it not leading the news. We were tired of it not being a part of the conversation around racial justice. We were like, 'What are we going to do next? What's the strategy?'" (Touré 2017, as cited in Hillstrom 2018: vii)

4   4 Regarding this connection between BLM and the media landscape in the USA, Rank/Pool (2023) write: "Our contemporary moment demonstrates that America's – and the world's – race question remains unsettled: from the eruption of Black Lives Matter protests in response to George Floyd's murder in May 2020 to the US government's Muslim ban and efforts to limit legal immigration and refugee claims to debates over Confederate statues and flags. Yet only recently has this increased racial awareness translated into the mainstream success of various television and movie projects that centre Black American experiences and are written, directed, or produced by Black people and star Black actors." (Rank/Pool 2023: 25)

with the political message of BLM. Upon its release, gamers of African descent worldwide praised the opportunity to identify with and play as a protagonist who shared their racial background, in a media landscape where non-White main characters remained the exception (Narcisse 2013; Marc 2018). In typical Ubisoft fashion, the game featured dialogues in Haitian creole, a detail that was praised by gamers of Afro-Caribbean heritage as a refreshing, moving and completely unexpected element. Narcisse (2013) stated in this regard: "Characters talk in Antillean Kreyol, the mosaic tongue made of French and West African words that I heard while growing up. But, mostly, it reminds me of going to church with my mother. It makes me happy and sad at the same time." (Narcisse 2013)

As we can gather from this introduction, this instalment of AC stands out on several levels: in its portrayal of a historical reality – the Haitian Maroon movement – that has been traditionally demonised in Western media; in its depiction of a charismatic Black protagonist at a time when Hollywood was largely ignoring and undervaluing this demographic; and in its powerful message that Black lives must be saved from slavery and white exploitation, resonating with the similar calls for justice and equality being voiced in the contemporary political arena.

In line with the above, ACFC raises several crucial questions about video games as platforms for historical and political engagement. What kinds of formal and aesthetic strategies allow a commercial video game to resist trivialising its historical material? How might procedural and spatial design be mobilised to reframe the experience of historical oppression and resistance, shifting the player's agency toward recognition and solidarity? And to what extent can such design choices elicit ethically meaningful responses from players, particularly in relation to a history as fraught as the transatlantic slave trade? These questions frame the critical orientation I adopt throughout this article, treating ACFC as a cultural artifact whose design opens a distinct mode of engaging with the past.

In the remainder of this article, I will explore ACFC through the lens of what I term 'the politics of history', focusing on how the game balances portraying historical events accurately while using them to critique and reflect on present-day inequalities rooted in that history. We must recognise, however, that ACFC is not a 'serious game' designed with educational or cultural goals in mind. Instead, it is part of a large entertainment franchise, where Ubisoft's primary objective is not to teach history but to entertain and profit by transforming historical content into a gamified experience, often at the risk of distorting historical facts to serve the goals of immersive gameplay. Yet, even within these limitations, the game demonstrates that mainstream video games can transcend the reinforcement of White male-dominated culture and the associated colonialist gaze that has long shaped the industry. Moreover, through empathic game design, video games such as this can act as mediators and creators of opportunities to symbolically challenge social stigma and erode that pervasive colonialist gaze.

2. History is our playground

Before turning to ACFC and its distinctive approach to historical content within the Ubisoft franchise, it is necessary to map the broader framework the series has established for engaging with the past. This will allow me to situate ACFC within Ubisoft's evolving representational strategies and to highlight the conceptual shift the game introduces, from what I will call a touristic model of historical simulation to a politically resonant engagement with the past.

As noted in the introduction, history has been a focus of game design almost since the inception of video games. Even a topic as complex and fraught as the gamification of societies based on slavery has been explored in games as far back as the early 8-bit era. Ubisoft, in particular, has made history a central theme in its blockbuster *Assassin's Creed* series. 'History is our playground' is a slogan used by Abstergo Industries, a fictional multinational conglomerate controlled by the Templars in the *Assassin's Creed* universe, whose main goal is to eradicate the order of Assassins, their arch-enemies.

Within the world of *Assassin's Creed*, history is accessed and experienced immersively through the Animus, a technology created by Abstergo Industries to combat the Assassins and recover ancient historical and prehistorical artifacts, with the ultimate goal of weaponizing them to achieve global domination. The Animus is also employed by the Assassins, enabling them to traverse history and battle the Templars across various regions and time periods. Building on this fictional premise, Ubisoft's *Assassin's Creed* franchise has developed a foundation for creating realistic, detailed, and immersive historical simulations spanning the ages, from ancient Egypt to Revolutionary France and Industrial England, from Classical Greece to 18th-century Havana and Saint-Domingue.

Several scholars (McCall 2022; Champion/Hiriart Vera 2023) have studied the educational potential of Ubisoft's historical simulations. Ubisoft, in turn, has supported this dialogue by collaborating with historians, linguists, and other experts to enhance accuracy in depicting regions, cultures, and historical eras. Since the release of *Assassin's Creed: Origins* (2017), the series has also included a Discovery Tour, curated by historians, enabling users to explore historical settings without combat or reliance on the game's fictional premises. However, even with the historically focused Discovery Tour series, the *Assassin's Creed* franchise retains a largely touristic approach to the past, as acknowledged by developers and scholars (Schwarz 2023). Referring to this approach as 'touristic' does not deny the games' historical or educational value but emphasises that: 1) historical content is subservient to gameplay and entertainment; 2) history is presented in a 'museum-like' or monumental fashion, with curated content focused on topics of popular appeal; and 3) the games disseminate established knowledge rather than generating new historical research.

This last point is particularly important, as Schwarz notes that in contemporary media, "the underlying assumption of every popular presentation of history [is that] what looks historical, that corresponds in its visualisations to the addressees' preconceived imagery, is historical" (Schwarz 2023: 171). As a result, period pieces – whether in TV series, films, or video games – tend to emphasise clothing, fashion accessories, monumental buildings, and historical settings to create an impression that aligns with the viewer's preconceived notions of what is 'historical'. The priority given to gameplay – understandably, as these are games – often leads to historical distortions, inaccuracies, anachronisms, and fictional content intertwining with the narrative. Advocates for using video games to teach history argue that these inaccuracies can provide valuable learning opportunities by encouraging learners' critical engagement (McCall 2022).

As we have seen up to this point, the way history is incorporated into the *Assassin's Creed* series is characterised by a balance between immersive entertainment and historical representation, with the latter often serving the former. While Ubisoft strives to introduce historical accuracy by involving experts and curating specific historical content, these elements remain largely subordinate to gameplay and narrative goals. The historical settings, presented in a 'museum-like' fashion, act as curated backdrops that prioritise popular appeal over academic rigour. This approach, which I refer to as 'touristic,' emphasises visual and surface-level accuracy – such as clothing, architecture, and settings – while allowing for historical liberties, distortions, and fictional content to enhance gameplay. Despite these shortcomings, the series has prompted discussions about its educational potential, with some arguing that the distortions and anachronisms could offer valuable opportunities for critical reflection on how history is portrayed in media (McCall 2022).

On the other hand, as I intend to demonstrate in this article, *Assassin's Creed* also takes an approach to historical representation that diverges from its museum-like treatment of the past, and is perceived through the contemporary political resonance of the events portrayed and the decisions developers make in how to depict them. In this sense, it is more accurate to speak of the politics of history rather than its digital monumentalisation. This take on history becomes most prominent in *AC: Freedom Cry,* as I will discuss next.

3. *AC: Freedom Cry:* Monumentalisation and the politics of history

A distinctive trademark of the AC franchise is its commitment to digitally modelling historical settings and simulating iconic monuments. From Jerusalem's Dome of the Rock in the first game to the intricately detailed 18th-century Paris in *AC: Unity*, each instalment lets players admire, explore, and climb (a key gameplay element) some of the most recognisable buildings of each location. However, developers sometimes take liberties with historical accuracy, such as including Havana's Catedral in ACBF before it was built. This highlights the franchise's focus on iconic structures for their visual and symbolic value, catering to a monumental approach to history aimed at a touristic gaze[5].

In contrast, ACFC stands out as one of the entries in the series most stripped of this monumental focus. As players explore the depiction of Saint-Domingue in the game, they are unlikely to grasp why the colony was considered one of the most profitable in the world, or why its principal city, Le Cap, earned the nickname 'The Paris of the Antilles'. Notably, Le Cap is absent from the game, which is set primarily in Port-au-Prince and its surrounding areas. Built in stone by colonial decree in 1753, Le Cap would have provided an ideal setting for the monumental recreations that the franchise is known for (Dubois 2009).

However, ACFC departs from the grand, touristic depiction of history, focusing instead on the political realities of slavery and rebellion while downplaying the architectural grandeur and colonial opulence typical of the series. This starkly contrasts with ACBF and ACL, where Havana and New Orleans are glamorised in ways absent from ACFC's Saint-Domingue. Port-au-Prince in this game lacks cultural significance and grandeur, featuring a relatively small environment with modest, utilitarian wooden structures instead of the stonework and monumental architecture of wealthier cities like Le Cap, which is omitted from the simulation. This lack of monumental design further impacts gameplay. The smaller scale makes walking, climbing, and parkouring less smooth. Players may often (as I did) find themselves unintentionally climbing buildings or taking wrong turns due to the limited space and reduced control when using buttons and joysticks.

While playing ACFC and engaging with its message and mechanics, it becomes evident that the abandonment of the touristic gaze is a deliberate design choice. However, this does not mean the game takes a deeper or more nuanced approach to pre-revolutionary Haiti. Instead, it adheres to Schwarz's concept of popular history: "what looks historical, that corresponds in its visualisations to

---

5   The concept of the 'tourist gaze' was developed by John Urry in *The Tourist Gaze: Leisure and Travel in Contemporary Societies* (1990), where he theorised that tourist experiences are shaped by socially constructed expectations and desires to encounter the 'other', often through a visual logic conditioned by class, gender, and cultural background. Drawing on Foucault's notion of the gaze, Urry framed tourism as a practice of power and representation, later expanding the concept in *The Tourist Gaze 2.0* (2002, with Jonas Larsen) to address critiques regarding its visual bias, lack of embodiment, and disregard for local agency. More recent discussions have introduced the 'reciprocal gaze' and other revisions attentive to mobility, sensory experience, and the performative nature of tourist interactions. In this article, I apply the concept of the 'touristic gaze' to describe Ubisoft's prevailing approach to history in the *Assassin's Creed* franchise, particularly in earlier titles such as *Black Flag*, where historical settings are curated as exotic, visually immersive, and consumable experiences designed to simulate a coherent but detached encounter with the past.

the addressees' preconceived imagery, is historical" (Schwarz 2023: 171). Depicting the golden age of Saint-Domingue in mid-18th century Le Cap would not only have been more historically accurate but also would have challenged gamers' assumptions about Haiti's history, particularly the notion of eternal poverty and lack of sophistication. Taking this route could have avoided another pitfall of the touristic gaze – the self-affirming presentation of historical knowledge – by offering a counterintuitive portrayal of Haiti's rich and complex colonial past, potentially sparking renewed interest in its history. However, this approach would have presented significant challenges. It risks glamorising colonial Haiti, built on the brutal exploitation of African slaves, or depicting the Maroon revolutionaries negatively, as destroyers of the colony's wealth, exactly the kind of racist narrative about Black Haitians propagated after their independence.

ACFC focuses instead on the politics of history, with its historical content stripped down to the essentials needed to support its political message. The Animus database, which provides expanded information on historical, cultural, and geographical content, includes fewer than two dozen entries. In contrast, ACBF has many more entries just for landmarks in Havana, Kingston, and Nassau. While ACBF is rich in historical references aimed at a touristic gaze – featuring famous Caribbean pirates like Calico Jack, Black Bart, Benjamin Hornigold, and Blackbeard, as well as missions in Caribbean forts and Mayan temples – ACFC narrows its scope. It centres its historical content primarily on slavery and the Maroon movement fighting to end it.

The story in ACFC is relatively minimalistic. Adewale, a charismatic character introduced in ACBF, begins the game shipwrecked near the coast of Saint-Domingue. Shortly before, he had obtained a parcel addressed to Bastienne Josèphe, a brothel owner in Port-au-Prince. Upon arriving in the city, Adewale becomes involved in a confrontation with an overseer chasing a runaway female slave. The first mission requires players to assist the woman by eliminating her pursuer. Afterward, players begin exploring Port-au-Prince, becoming aware of its dangers – such as jailers scattered throughout the city searching for runaway slaves – and witnessing the abuse inflicted by overseers and colonists. The game focuses on specific scenarios where players can intervene: stopping overseers from capturing runaway slaves, rescuing slaves being punished, freeing those kept in cages or sold in the market, and helping wounded slaves to safety. Whenever Adewale succeeds in freeing a slave, he conveys a political message about the need to unite and show solidarity 'to our brothers and sisters'. As the game progresses, Adewale's mission will involve the recruitment of liberated slaves for the Maroon movement.

While completing these initial quests, the connection between this simulation and the gameplay style of ACBF quickly becomes clear; after all, this game is essentially an expansion pack of the latter. As noted, Port-au-Prince lacks the glamour of ACBF's Havana or ACL's New Orleans but closely resembles Nassau as portrayed in ACBF. Moreover, the quest to liberate slaves is essentially an adaptation of Kenway's mission in Nassau to free captured pirates and recruit them for his crew.

In making this connection, we can start to notice ways in which ACBF and its expansion are ideologically related: both lean heavily on the topic of freedom, but while ACBF centers on the romanticised freedom of pirates who roam the oceans on the margins of regulated society in the eighteenth century, ACFC addresses the less romantic, urgent, and more contemporarily relevant freedom of individuals of African descent. Despite using the same aesthetics (and likely the same graphics and assets) to portray Nassau and Port-au-Prince, the contrast between the touristic approach to piracy in ACBF and the political approach to Black oppression in ACFC could not be starker. The two games, taken together, reveal Ubisoft's dual approach to the curation of historical content within the Caribbean timeline of the Assassin's Creed saga. ACBF invokes the monumental through its celebration of piracy's romanticised mythos, while ACFC resituates the narrative within the political, foregrounding the urgent realities of Black resistance and liberation.

In the following sections, I will conduct a more detailed close reading of the game's narrative and its procedurally encoded messages, examining the type of reception it anticipates from players. This reception can be characterised as a quest to encode empathic gameplay rooted in an ethical framework that more closely aligns with the value systems of a contemporary progressive society, reflecting the political ethos of the Obama years when the game was released

4. The fragility of hope: Designing empathic gameplay in ACFC

A game that seeks to simulate the transatlantic slave system with a realistic portrayal of the slave trade and the dehumanisation of people of African descent faces significant risks. This challenge was starkly illustrated by *Playing History: Slave Trade* (PHST), released in 2013 by Copenhagen-based Serious Games Interactive. Unlike ACFC, PHST explicitly aimed to educate players about the brutal realities of the slave trade but became highly controversial due to one particular level. In this segment, players were tasked with filling a slave ship using a Tetris-like mechanic, arranging human cargo to optimise space. This design choice, which trivialised the horrors of slavery into a puzzle-solving exercise, provoked widespread backlash and accusations of insensitivity (Sepinwall 2021: 186–188).

By contrast, ACFC presents the same historically accurate details about the inhuman conditions endured by African slaves aboard ships. However, both narratively and through its procedurally encoded messages, the game adopts a more tactful approach, preserving the subject's gravity while still delivering an engaging and, as expected from a commercial video game, enjoyable experience.

Part of the success of ACFC can be attributed to what I will refer to as empathic game design. The mechanisms by which empathy operates in media consumption have recently been the subject of extensive scholarly investigation (Tettegah/

Espelage 2015; Muriel/Crawford 2018; McStay 2018). A working definition of empathy frames it as "an emotional response that stems from another's emotional state or condition and is congruent with the other's emotional state or condition" (cf. Happ/Melzer 2014: 3-4). In social interactions and cultural products, empathy can be present both as a psychological characteristic of the individual – namely, the propensity of the individual to be more or less empathic – and as a demand potentially arising from a situation, that is, the circumstances that may prompt an empathic response (Happ/Melzer 2014). Research has shown that empathy, along with the pro-social behaviour it fosters, is not solely an inherent trait of certain individuals but can also be learned and cultivated[6]. Exercises in perspective-taking can enhance competence in cognitive empathy and lead to such results.

This explains the increasing attention in video game research and design to the medium's potential for fostering empathy in players (Belmanand/Flanagan 2010; Bleiker 2018: 115–120; Happ/Melzer 2014). Unlike traditional media, video games allow players to control characters, make decisions on their behalf, and directly experience the consequences of those decisions. This unique encoding of character agency enhances perspective-taking, which may, in turn, evoke stronger empathic responses to the situations presented.

A video game on transatlantic slavery and the Maroon revolt, such as ACFC, provides a unique platform to explore this framework. Given the emotionally charged and morally complex nature of these themes, the game's success depends on its ability to evoke empathy, fostering identification with the subject matter. Empathy serves as a crucial moderator (Happ/Melzer 2014), enabling players to engage with the harsh realities of slavery in a way that promotes understanding rather than desensitisation. The political approach to the history of slavery analysed here – resonating with contemporary issues like the unfulfilled promises of Black emancipation highlighted by movements such as BLM – relies on this empathic connection to make ACFC a powerful medium for social commentary. With this in mind, I will next evaluate the game's success in achieving this through a close reading of its narrative and procedurally encoded messages, alongside an analysis of players' responses.

Some elements of ACFC contributing to its empathic design are straightforward yet uncommon in the video game industry, such as the inclusion of a playable hero of African descent, a man born into slavery who becomes a liberator. By embodying this protagonist and controlling his agency, players step into the role of a historically inspired figure fighting to end slavery. The protagonist's 'fighting' traits are another significant aspect. As part of the *Assassin's Creed* franchise, ACFC includes combat and assassination missions, aligning with conventions

---

[6] Happ and Melzer (2014) observe in this regard: "Even though empathy may have a strong evolutionary basis, empathy as a personality trait can be learned or promoted in trainings." (4)

often criticised for glorifying violence. However, in this context, the fighting feels more justified, given the historical subject and cause.

Adewale is not relatable for pacifist qualities often attributed to leaders of oppressed minorities, like Gandhi, Nelson Mandela, or Martin Luther King Jr. Instead, he is a fierce fighter wielding a machete – imagery historically used to demonise Black men, particularly those who fought to free slaves in Haiti. In ACFC, this stigma is reversed, casting Adewale as a champion of Black rights with whom players can identify, as evidenced by their reactions to the game. This identification is encoded into the narrative and missions, shaping how players connect with the character.

At the beginning of the game, Adewale is introduced as just another assassin in the series, a character players recognise from ACBF, where he served as the charismatic quartermaster to the main playable character, Edward Kenway. ACFC charts Adewale's transformation from a loyal member of the Brotherhood to a Maroon leader, as he personally witnesses the abuses inflicted on slaves. This narrative arc not only deepens his character but also serves as a blueprint for modelling the same expected internal journey in players, encouraging them to grapple with the brutality of slavery and the fight for freedom as the game progresses. At both the narrative and game mechanics levels, the game strives to facilitate this identification with the subject matter by including historical and gameplay elements that support this goal and removing aspects – such as its trademark interest in the monumental – that might detract from its political objectives.

Visual design contributes powerfully to this reorientation. Unlike previous titles in the franchise, ACFC eschews the aesthetic grandeur typically associated with the Assassin's Creed brand. The game's visual language is instead deliberately sparse: buildings in Port-au-Prince are weathered and wooden, streets dusty and unpaved. The effect is neither majestic nor picturesque but grimly utilitarian. This visual realism grounds the story and supports its political aims. Players are offered few picturesque moments and are not encouraged to marvel at the skyline or architecture. Instead, they are confronted with a geography of suffering: slaves in chains, overseers on patrol, bodies hanging as warnings. These repeated and understated visual cues accumulate slowly, embedding players within a world of systemic violence that does not require spectacular set pieces to feel oppressive.

This visual paring-down continues in the design of plantation spaces, where the isometric rows of sugarcane and the symmetrical placement of slave quarters produce a spatial logic of control. Visual cues here are both symbolic and disciplinary: overseers' lines of sight, visible through the game's mechanics, mirror the panoptic structure of surveillance. Even the silhouettes of watchtowers function not as scenic details but as instruments of domination. Adewale's movement through these spaces, his stealth, his forced crouching, constitutes a kinetic mapping of resistance, shaped by and responding to these architectural constraints.

As I have already outlined, ACFC begins off the coast of Saint-Domingue, thirteen years after the events of ACBF. The initial campaign involves quests within the city of Port-au-Prince, but the game soon expands to include a political campaign beyond these boundaries. After meeting Bastienne Josèphe and forming a tense partnership marked both by distrust and moments of connection, Adewale is tasked with delivering a message to one of her contacts at a plantation outside the city. At the plantation, players can free up to 40 slaves by eliminating 20 overseers enforcing the order. The quest features the franchise's characteristic mix of stealth and efficient killing, with stealth being particularly crucial in ACFC. Direct confrontations can have catastrophic outcomes: if an overseer sounds the plantation bell, reinforcements are alerted, and the mission fails. Even if all bells are preemptively disabled, open combat risks overseers retaliating by killing the enslaved individuals, adding a dire sense of urgency and ethical resonance to the mission.

In this way, the game seeks to foster empathic gameplay, encouraging players to connect emotionally. Protecting enslaved individuals from overseers – and often preventing their deaths – is the central mission for Adewale and the players controlling him. In his interactions with enslaved NPCs, Adewale – and by extension, the players – encounters the Gramscian concept of cultural hegemony (Schwarzmantel 2014) and internalised oppression (Sullivan/Cross 2016). Enslaved NPCs often respond to Adewale's actions with fear and apprehension rather than hope. "*Tu vas nous faire tous tuer*" ("You're gonna get us all killed") is a typical outcry whenever he kills an overseer in their presence[7]. This reaction does not stem from a lack of desire for freedom; their willingness to flee and hide after their liberation demonstrates otherwise. Instead, it reflects doubt in Adewale's ability and their own capacity to resist oppression. In this way, the NPCs embody internalised acceptance of oppression while offering a meta-commentary on the mission undertaken by the players.

The plantation liberation quest expands as Adewale returns to the Caribbean waters. As captain of the *Experto Crede*, he must intercept slave ships to liberate their captives. Building on the naval combat mechanics introduced in *Assassin's Creed III* and refined in ACBF, ACFC tasks players with defeating the fleet escorting the slave ships while keeping the slave ship unharmed for boarding and liberation by Adewale and his crew. Like plantation and city missions, naval combat requires strategy: players must balance aggressive attacks to defeat enemies with the need to ensure the safety of the enslaved individuals aboard.

The game does not shy away from confronting players with brutal scenes of unavoidable failure, intentionally designed to evoke feelings of impotence and frustration. This is particularly evident in one of the final missions, *Memory*

---

[7] Another slave NPC tells Adewale when she sees him in the plantation: "*On n'a pas besoin de toi. Va-t'en!*" ("We don't need you. Go away!").

8: *Down with the Ship.* By this point, Adewale's campaign to liberate slaves and strengthen the Maroon movement has drawn the attention of French colonists and their leadership, represented by Gouverneur de Fayet, a character based on a historical figure. As a result, enslaved individuals are facing heightened surveillance and harsher punishments. In a cutscene, Bastienne Josèphe warns Adewale of the unintended consequences of his actions, pointing to the suffering endured by those he cannot protect: "You help the ones you freed, but at what cost to the ones you can't?" This dialogue sets the stage for one of the game's most impactful missions. During the ensuing sequence, Adewale attempts – and fails – to liberate slaves aboard a ship, with his failure depicted in a particularly graphic and harrowing manner.

While engaging the escorting ships in what begins as a routine naval battle, Adewale notices one vessel attempting to sink the ship carrying enslaved individuals. In a sudden escalation, Adewale – and the players – must act swiftly to defeat the enemy and protect the 'precious cargo'. Upon boarding the ship, Adewale encounters a grim scene: the vessel is sinking, and the enslaved remain chained below deck in cramped, inhumane quarters. Many appear dead or unconscious as water pours in, while others cry out in pain and stretch their arms toward Adewale, pleading for rescue. Explosions rock the ship, and the upper deck collapses, signalling its inevitable submersion with most captives still trapped aboard.

As the ship begins to capsize, roughly three minutes into the mission, bodies tumble into the water, and Adewale is pulled into the flooded interior, surrounded by the submerged dead and dying. The emotional impact is heightened by the musical backdrop: a stirring variation of *The Root*, the game's theme song, performed by Haitian musician Alexis Monvelyno in collaboration with composer Olivier Derivière. Inspired by the traditional Haitian song *Fyè Aleman Lèmiso Batala (Pierre Allemand, the Spirit of Battle)*, the haunting tones amplify the devastation, immersing players further in Adewale's harrowing reality.

By now, it should be clear that ACFC prioritises fostering empathic gameplay through a straightforward and unembellished social and political narrative. The historical content is intentionally stripped of ornamental details that might defamiliarise players from the events by introducing elements foreign to their present-day experience. This reduction of exoticizing elements refocuses players' attention, enabling a more subjective and emotionally resonant engagement with the story. Additionally, many historical and geographical aspects that might otherwise seem unfamiliar – such as navigating the Caribbean's network of islands or understanding the technology of colonial-era vessels, from schooners to Man O' Wars – were already introduced in ACBF, preparing players for the expansion's more politically charged narrative.

At the storytelling level, ACFC adopts a similarly minimalist approach, focusing on essential narrative elements. The cast features archetypal characters who drive the plot and reinforce its themes. Adewale, the hero, is joined by Bastienne Josèphe, a pragmatic recruiter who helps organise the Maroon resistance, and Augustin Dieufort, a mentor and leader of the Maroons. Opposing them is Governor Pierre de Fayet, embodying colonial cruelty and institutionalised violence. Supporting these primary characters is a small but symbolically significant group of NPCs, each representing distinct sociological archetypes.

Among the enslaved NPCs, players encounter runaways who embody defiance and the yearning for freedom, as well as those depicted in distress or suffering brutal abuse, reinforcing the stakes of Adewale's mission. Others remain compliant with the hegemonic order, working the plantations and discouraging insurgency out of fear or resignation. These figures add depth to the sociopolitical landscape, illustrating the varied ways individuals navigate oppressive systems. On the opposing side are cruel overseers, ranging from standard antagonists to tougher 'bosses' in line with classic game design, where enemies become harder to defeat at higher levels. This progression extends to naval combat, with enemy vessels growing increasingly difficult to disable as the game advances.

As a result, the game's central narrative and core ideologeme are rooted more in sociology and politics than history. Its focus on the fragility of hope – powerfully depicted in the sequence *Down with the Ship* – highlights resistance and oppression's sociopolitical dynamics within a non-diachronic timeline, rather than offering a detailed historical reconstruction. This approach underscores the precariousness of liberation movements, where defiance coexists with systemic violence and vulnerability. By centring these themes, ACFC grounds itself in the politics of history, departing from the monumental approach typical of Ubisoft's saga.

An examination of how ACFC has been received by the gaming community will further offer an opportunity to explore both its political messaging and the relative effectiveness of its empathic game design. To achieve this, I will finally analyse player-reported reactions, including critical reviews, forum discussions on ACFC, and responses from YouTubers streaming their gameplay. This approach aims to provide a clearer understanding of how these dynamics are experienced, interpreted, and observed through the lens of player accounts.

5. The Sounds of liberty: Players's reactions to ACFC

Sepinwall (2021) has highlighted some of the evidence available regarding the impact ACFC had on players. She cites in this regard a forum created by Ubisoft in response to players' interest in learning more about the Haitian Maroon movement and revolution:

"Players were so enthusiastic about the Animus information that Ubisoft added further information on the Revolution in *AC Initiates*, a section of Ubisoft's website with bonus content. This content fictionalised portions of the Haitian Revolution, appropriating it into the *Freedom Cry* storyline. While the forum is defunct, fan pages reproducing this material indicate that it identified Toussaint himself as a member of the Brotherhood of Assassins." (Sepinwall 2021: 201)

Sepinwall observes that, "while this information is fictional, it did introduce more Haitian Revolution content to AC players", while also creating "the possibility that players might seek to learn about these real figures from more reliable sources" (ibid: 203). Player reactions to the game, including specific missions and scenes, can be further explored through their comments on various internet forums, social media platforms[8], and streamed 'walkthroughs' of the game's missions.

Reactions to the liberation of slaves in a plantation mission are particularly telling, capturing players in real time as they vocally identify with the main character and his quest while reflecting on the resonance of the portrayed situations. After completing the mission, ShakeDown2012, a YouTuber of African descent, remarked: "Now, there is nobody else here [at the plantation]. I like this. I did this!" (*Walkthrough #2*) Earlier, while sabotaging one of the bells, he commented: "Let's sabotage the bell so that they can't call for help, [if] they notice a Black man with a hood on."[9] (ibid) Frequently, after completing in-game quests, ShakeDown2012 expresses pride in his accomplishments. In another stream, he declared: "You saw what I just did: liberating people, liberating my brothers!" (*Walkthrough #1*) Notably, in his first stream of the game, he mentioned his intention to post his initial walkthrough of ACFC a day earlier, coinciding with Martin Luther King Jr. Day to commemorate the holiday.

Terrence Dearman, another YouTuber, praised the traditional Haitian songs featured during the plantation liberation mission: "Ubisoft really knew what they were doing when they put this stuff together. It is very accurate," he stated (*Walkthrough Part 3*). Dearman further used the opportunity of visiting a Maroon hideout to educate his viewers on the Marocn movement, its goals, and its activities (ibid). After completing the plantation liberation mission, another YouTuber, GameRiot also declared: "I freed loads of people! Awesome!" before quickly acknowledging the sensitivity of the subject, adding: "It is a very tricky subject to do in a game." (*Walkthrough Part 2*) Similarly, Cata or Shane, another streamer, modelled a style of gameplay that highlighted the educational potential of ACFC. She frequently

---

8   Sepinwall (2021) notes that "on Twitter, those who represent themselves as subscribers of colour expressed their amazement about playing the game – or just learning about its existence." (205) Readers interested in this topic and the pitfalls of the game mechanics – such as offering monetary rewards for liberation campaigns – are encouraged to consult her book.

9   Let us recall that the BLM movement was largely sparked by the killing of Black teenager Trayvon Martin by vigilante George Zimmerman, who was acquitted at trial. Though innocent and unarmed, Martin was perceived by Zimmerman as 'up to no good' based solely on his skin colour and hoodie. In this context, playing as a hooded character can carry strong relatable connotations for Black players, as suggested by ShakeDown2012.

paused to read entries from the Animus database, relaying the historical content aloud and elaborating based on her prior knowledge of the subject. The importance of completing liberation missions stealthily, to prevent overseers from killing enslaved NPCs, was emphasised by several other streamers (MysteriousJG, FightinCowboy). Some commentators even criticised certain YouTubers for engaging with overseers in ways that led to unnecessary deaths of enslaved NPCs, underscoring the ethical considerations embedded in the gameplay[10].

Players' reactions to the mission 'Down with the Ship' further reveal their empathic engagement with the story and its overarching message. In this quest, as previously described, Adewale is unable to free the enslaved individuals aboard a slave ship before it capsizes, leaving most to perish while still chained to the vessel. Numerous streamers expressed their real-time consternation and devastation during these scenes, while viewers' comments further illustrate the reception of this fictional recreation of a possible historical scenario by Ubisoft:

"Running out of the ship while all the slaves are trapped and yelling feels like hell itself, especially because you know stuff like this happened constantly back then," wrote @haitharu in response to the Assassin's Creed Series channel's stream of the mission (Assassin's Creed Series 2016).

"So sad climbing up the ship, surround[ed] by slaves hanging from their legs, screaming for help as the water rises, knowing you cannot save them..." wrote @danhughes581 (Ibid).

"This is one of the saddest moments in the series. No matter how fast you are, you can't save them all. I stopped playing *Rogue* when I was told I had to kill Adewale – I just couldn't, not after playing this," added @isaiahcollins3451 (Ibid).

"This scene breaks you because not only can you not save them all, and you have to watch them die, but in some sense, it was all your fault for not listening to the Lady's warning," reflected @kokotree9226 (Ibid).

"I have played all the *Assassin's Creed* games many times, but this mission made me so sad and mad—how evil people were to enslave human beings like this. I don't think there has been any slavery as evil as the Atlantic slave trade," stated @Yasine13 (Ibid).

The quoted evidence highlights players' strong empathic responses to the mission 'Down with the Ship'. Empathy is demonstrated through players' ability to connect emotionally with the suffering depicted in the scene, despite its fictional nature.

---

10   After completing the mission with a count of only 3 out of 30 slaves still alive, the player streaming in the channel AFGuiderHD, received a comment asking them to: "Plz be more stealthy."

Comments like those from @haitharu and @danhughes581 reflect players' visceral reactions, as they vividly describe the helplessness and horror of being unable to save the enslaved characters. These reactions show not just engagement with the game mechanics but an alignment with the historical gravity of the narrative.

At a linguistic level, these reactions reveal a striking level of identification with the game's protagonist, Adewale. Players frequently use second-person pronouns such as "you cannot save" (from @danhughes581) and "it was your fault" (from @kokotree9226), positioning themselves as the subject directly responsible for the outcomes in the scene. This choice of language suggests a deep immersion in the narrative, where players do not distance themselves by attributing responsibility solely to the game developers or the fictional character they control. Instead, they claim ownership of the moral and emotional weight of the mission's events, internalising the consequences as though they were personally involved.

Moreover, the players' use of active voice – "you have to watch them die" and "I stopped playing" – demonstrates a sense of agency within the game's framework, even in a scenario where the outcome is predetermined by the developers. Phrases such as "I like this, I did this" (from ShakeDown2012) and "I freed loads of people" (from GameRiot) earlier in the game reinforce this active, first-person engagement. This linguistic structure indicates that players interpret the game's events as a reflection of their own choices and actions, even when constrained by the game's narrative design. The emotional intensity is further amplified by the use of vivid sensory language, such as "trapped and yelling feels like hell itself" and "screaming for help as the water rises". These descriptions transform the abstract experience of playing a game into a tangible, almost physical ordeal, blurring the line between fictional empathy and real-world emotional engagement. Similarly, moral evaluations such as "how evil people were to enslave human beings like this" (from @Yasine13) reveal how players expand their empathic reactions to include broader historical and ethical reflections.

The linguistic analysis underscores how ACFC enables a unique intersection between narrative immersion and personal identification. By constructing responses that place themselves as moral agents within the story, players demonstrate how empathy in interactive media is both a linguistic and emotional phenomenon, driven by the interplay of narrative design and player agency. This dynamic not only deepens their engagement with the game but also fosters a more profound connection to the historical and ethical issues it portrays.

All in all, ACFC offers a compelling demonstration of how a commercial video game can navigate the complexities of historical representation on a challenging subject while fostering deep empathic engagement with its players. As mentioned, Ubisoft stepped out of its comfort zone with this game, abandoning its tried-and-true monumental approach to the past in favour of a more politically charged narrative. Through its empathic game design, the game transforms historical content into a visceral, emotionally resonant experience, as evidenced by the player reactions discussed, reflecting a blend of moral reflection, personal identification, and sensory immersion. The minimalist yet powerful storytelling, combined with deliberate gameplay mechanics, guides players to confront both the horrors of oppression and the precariousness of liberation, framing the experience in sociopolitical rather than purely historical terms. In doing so, ACFC not only pushes the boundaries of what interactive media can achieve but also sparks broader conversations about the potential of video games to serve as platforms for critical reflection on enduring issues of inequality and resistance.

6. Conclusion: From archive to encounter: Empathic design and the politics of history in *Freedom Cry*

As demonstrated in the preceding discussion, players' responses to missions such as the plantation liberations, the assistance of runaways, and especially 'Down with the Ship', reveal the emotionally charged and ethically engaged gameplay that *Assassin's Creed: Freedom Cry* makes possible. These responses reflect more than momentary reactions to dramatic scenes; they illustrate a sustained form of narrative immersion in which players assume a position of moral agency within a historical context. In walkthroughs and comment threads, players consistently adopt a language of identification that blurs the boundary between themselves and Adewale. Comments such as "I like this. I did this", "liberating my brothers", or "it was all my fault" indicate an affective and ethical alignment with the protagonist's perspective. The use of first-person declarations and second-person constructions, especially those that assign responsibility, demonstrates a marked investment in the lives at stake and suggests that the emotional weight of the missions was internalised as a personal experience rather than simply observed from a distance.

The structure of the game, including its use of stealth mechanics and mission design, encourages this affective alignment by making liberation contingent upon careful, deliberate action. The moral implications of failure, foregrounded by reactions to overseer violence or to catastrophic missions like 'Down with the Ship', intensify the player's sense of involvement and complicity. What emerges is a model of play that combines narrative direction with experiential affect, where the fictional environment becomes a space for reckoning with real historical violence.

In that sense, *Freedom Cry* does not simply portray a revolutionary struggle, it channels it through the player's own agency. Empathy here is not imposed by cutscenes or moralising dialogue, but arises through gameplay choices, failed rescues, successful liberations, and affective responses voiced aloud by players and documented by their audiences. The history in *Freedom Cry* becomes playable not because it grants full control or historical verisimilitude, but because it constructs a field of consequences where ethical urgency and narrative meaning intersect.

Ubisoft's choice to shift from monumental reconstructions of history toward a more intimate, politically charged depiction of resistance creates a productive tension in *Freedom Cry*. Through sparse but pointed storytelling, atmospheric sound design, and an emphasis on marginalised perspectives, the game reconfigures its historical material into a catalyst for player-driven ethical reflection. This design opens a space where empathy operates through identification and memory, encouraging players to engage critically with the legacies of slavery, rebellion, and historical erasure.

References

AFGuiderHD (21 December 2013): "Assassin's Creed 4 Freedom Cry DLC - Plantation Liberation - Lagon Nord" [Video]. YouTube. https://www.youtube.com/watch?v=zUStyv4W3TY&t=206s.

Assassin's Creed Series (8 September 2016): "Assassin's Creed: Freedom Cry - Memory 08 - Down With the Ship" [Video]. YouTube. https://www.youtube.com/watch?v=vzae6tR3Gik.

Belman, J./Flanagan, M. (2010): "Designing Games to Foster Empathy." Cognitive Technology 14(2), pp. 5–15.

Bleiker, R. (2018): Visual Global Politics. 1st ed. London/New York: Routledge.

Brock, A. (2011): "'When Keeping It Real Goes Wrong': Resident Evil 5, Racial Representation, and Gamers." Games and Culture 6(5), pp. 429–452. https://doi.org/10.1177/1555412011402676.

Cata or Shane (15 April 2023): "Assassin's Creed: Freedom Cry - Episode 2 (A Common Enemy)" [Video]. YouTube. https://www.youtube.com/watch?v=XRhImgLIuUY&t=599s.

Champion, E./Hiriart Vera, J. F. (eds.) (2024): Assassin's Creed in the Classroom: History's Playground or a Stab in the Dark? Berlin/Boston: De Gruyter Oldenbourg.

Cleger, O. (2016): "Why Videogames: Ludology Meets Latino Studies." In: F. L. Aldama (ed.) The Routledge Companion to Latina/o Popular Culture. London/New York: Routledge, pp. 87–100.

Coktel Vision (1988): Freedom: Rebels in the Darkness [Video game]. Coktel Vision.

Dearman, T. (18 December 2013): "Assassins Creed 4 Black Flag DLC - Freedom Cry Walkthrough Part 3 - Laying the first brick" [Video]. YouTube. https://www.youtube.com/watch?v=wBhahNPGb4w&t=884s.

Dubois, L. (2009): Avengers of the New World. Cambridge: Belknap Press.

Feagin, J. R. (2020): The White Racial Frame: Centuries of Racial Framing and Counter-Framing. London/New York: Routledge.

FightinCowboy (20 December 2013): "Assassin's Creed 4 Black Flag Freedom Cry - Gameplay Walkthrough Part 3: Laying the First Brick" [Video]. YouTube. https://www.youtube.com/watch?v=Rn3IBjzcDew&t=386s.

GameRiot (17 December 2013): "Assassin's Creed 4 Black Flag - Freedom Cry Walkthrough Part 2 Liberating" [Video]. YouTube. https://www.youtube.com/watch?v=O_VURhcsq8c&t=398s.

Goren, L. J./Carnes, N. (2023): "An Introduction to the Politics of the Marvel Cinematic Universe." In: L. J. Goren/N. Carnes (eds.), The Politics of the Marvel Cinematic Universe. Lawrence: University Press of Kansas, pp. 1–18.

Hammar, E. L. (2017): "Counter-hegemonic Commemorative Play: Marginalized Pasts and the Politics of Memory in the Digital Game Assassin's Creed: Freedom Cry." Rethinking History 21(3), pp. 372–395.

Happ, C./Melzer, A. (2014): Empathy and Violent Video Games. Basingstoke: Palgrave Macmillan.

Hillstrom, L. C. (2018): Black Lives Matter: From a Moment to a Movement. Santa Barbara: Greenwood.

Hoermann, R. (2016): "'A Very Hell of Horrors'? The Haitian Revolution and the Early Transatlantic Haitian Gothic." Slavery & Abolition 37(1), pp. 183–205.

Lauro, S. J. (2020): Kill the Overseer! The Gamification of Slave Resistance. Minneapolis: University of Minnesota Press.

Marc, B. (6 December 2018): "Gaming While Black: Has Red Dead Redemption 2 Shifted a Toxic Culture?" In: The Hollywood Reporter. Retrieved from https://www.hollywoodreporter.com/news/general-news/subversive-racial-commentary-red-dead-redemption-2-1165989/.

McCall, J. (2022): Gaming the Past: Using Video Games to Teach Secondary History. London/New York: Routledge.

McStay, A. (2018): Emotional AI: The Rise of Empathic Media. London: SAGE Publications Ltd.

Muriel, D./Crawford, G. (2018): Video Games as Culture. London/New York: Routledge.

MysteriousJG (11 January 2021): "Let's Play Assassin's Creed IV Black Flag Part 121: Freedom Cry Part 6" [Video]. YouTube. https://www.youtube.com/watch?v=cbLtjkW8aoY .

Narcisse, E. (19 December 2013): "A Game That Showed Me My Own Black History." Retrieved from https://kotaku.com/a-game-that-showed-me-my-own-black-history-1486643518.

Rank, A./Pool, H. (2023): "Building Worlds: Three Paths Toward Racial Justice in Black Panther." In: L. J. Goren/N. Carnes (eds.), The Politics of the Marvel Cinematic Universe. Lawrence: University Press of Kansas, pp. 19–35.

ShakeDown2012 (January 19 2016): "Assassin's Creed Freedom Cry Walkthrough Gameplay #1 A Common Enemy" [Video]. YouTube. https://www.youtube.com/watch?v=e9-j4J71ScE

ShakeDown2012 (January 20 2016): "Assassin's Creed Freedom Cry Walkthrough Gameplay #2 Plantation Liberation" [Video]. YouTube. https://www.youtube.com/watch?v=LcgssLylFbI

Schwarz, A. (2023): "Discovering the Past as a Virtual Foreign Country: Assassin's Creed as Historical Tourism." In: J. F. Hiriart Vera/E. Champion (eds.), Assassin's Creed in the Classroom: History's Playground or a Stab in the Dark? Berlin/Boston: De Gruyter, pp. 159–187.

Schwarzmantel, J. (2014): The Routledge Guidebook to Gramsci's Prison Notebooks. London/New York: Routledge.

Sepinwall, A. G. (2021): Slave Revolt on Screen. Jackson: University Press of Mississippi.

Sullivan, J./Cross, W. (2016): Meaning-Making, Internalized Racism, and African American Identity. Albany: SUNY Press.

Szaban, D./Szymczyk, M./Kida, M. (2025). Designing and Implementing Public Policy of Contemporary Polish Society. V&R Unipress.

Taylor, K.-Y. (2016): From #BlackLivesMatter to Black Liberation. Chicago: Haymarket Books.

Tettegah, S./Espelage, D. (2015): Emotions, Technology, and Behaviors. San Diego, CA: Academic Press.

Thomas, M./Wendling, M. (15 September 2024): "Trump Repeats Baseless Claim About Haitian Immigrants Eating Pets". BBC News. Retrieved from https://www.bbc.com/news/articles/c77l28myezko.

Trouillot, M.-R. (1995): Silencing the Past: Power and the Production of History. Beacon Press.

Ubisoft (2012): Assassin's Creed III: Liberation [Video game]. Ubisoft.

Ubisoft (2013): Assassin's Creed: Freedom Cry [Video game]. Ubisoft.

Ubisoft (2013): Assassin's Creed IV: Black Flag [Video game]. Ubisoft.

Ubisoft (2017): Assassin's Creed: Origins [Video game]. Ubisoft.

Urry, J. (1990): The Tourist Gaze: Leisure and Travel in Contemporary Societies. London: Sage.

Urry, J./Larsen, J. (2011): The tourist gaze 3.0. 3rd ed. London: SAGE Publications.

Wang, Z. (2012): Never Forget National Humiliation: Historical Memory in Chinese Politics and Foreign Relations. Columbia University Press.

Williams, D./Martins, N./Consalvo, M./Ivory, J. D. (2009): "The Virtual Census: Representations of Gender, Race, and Age in Video Games." New Media & Society 11(5), pp. 815–834.

Yuen, N. W. (2016): Reel Inequality. New Brunswick: Rutgers University Press.

Historical Empathy and Player Agency in Computer Roleplaying Games
Kingdom Come: Deliverance and Pentiment

Robert Houghton

Abstract

This chapter contends that Computer Roleplaying Games (commonly CRPGs) provide substantial potential for the development of historical empathy both within and outwith the classroom. In doing so, it adapts and develops a large body of existing scholarship which demonstrates the ability of face to face roleplaying activities to inspire historical empathy amongst their players and recent work highlighting the potential of realist games (such as Assassin's Creed and Valiant Hearts) to promote similar empathetic experiences. Through case studies of the roleplaying games Kingdom Come: Deliverance and Pentiment, this piece acknowledges the similarities in empathetic qualities between roleplaying games and realist games, but argues that the nature and conventions around CRPGs – particularly the agency they grant their players and the impact this has on avatar embodiment and engagement with the game world – permit them to provide distinct and powerful opportunities for the nurturing of historical empathy. Ultimately, the chapter argues that roleplaying games need to be considered more seriously within historical games studies, and can provide important alternative approaches to history from those prominent within realist and conceptual games.

Keywords

roleplaying, games, CRPGs, historical empathy, player agency

Roleplaying games, empathy, and history

Although the label 'roleplaying game' is well established within common parlance and abounds in descriptions of both digital and analogue games, as is the case with any genre of game, the boundaries and definitions of roleplaying games are porous and often disputed. As José Zagal and Sebastian Deterding demonstrate, the definition of 'roleplaying game' is further muddied by the often stark differences between tabletop, Live Action (LARPs), single player CRPGs, and Massively

Multiplayer Online Roleplaying Games (MMORPGs) (Zagal/Deterding 2018), not to mention further differentiation between 'Western' RPGs and Japanese (JRPGs). The broadest interpretation – that roleplaying refers to any game where the player(s) take on the role of a character other than themselves – is applicable to all but the most abstract of games and therefore of very little use. Restricting 'roleplaying games' to games where roleplaying and interpersonal relationships are central to play can be more useful, but is still hugely subjective and can be stretched to include a significant number of games which would not typically be classified as roleplaying. We can profitably narrow this range of games through Jonne Arjoranta's (2011: 14) proposed definition – itself an evolution of that of Michael Hitchens and Anders Drachen (Hitchens/Drachen 2008) – presenting roleplaying games as those where the player holds "shared narrative power" (with the ability to substantially alter the narrative) and has "varying modes of interaction with the game world" (facilitating different, if limited, ways of approaching problems posed within the game). The definition is one of agency: roleplaying games allow the player to change their worlds in meaningful ways.

This definition of roleplaying games (adopting the role and personality of another character with the ability to meaningfully impact the game world through various means) is undoubtedly still contestable, certainly still woolly, and should not be seen as a rigid and impermeable boundary between games which facilitate roleplaying and those which do not. However, this model allows the distinction between computer roleplaying games such as *Kingdom Come: Deliverance* (Warhorse Studios 2018) and games with roleplaying elements such as most of the *Assassin's Creed* (Ubisoft 2007-2025) series. The former focuses on characterisation and character development, allows players to exert a meaningful impact on the game world, and provides various means to approach issues. The latter sidelines characterisation and limits character development, follows a very linear storyline, and largely restricts player's solutions to in game problems to violence. There is plenty of blurring here: *Assassin's Creed: Odyssey* (Ubisoft, 2018) and, to a lesser extent, *Valhalla* (Ubisoft, 2020) and *Shadows* (Ubisoft, 2025) featured more opportunity for the player to meaningfully influence their world for example. Nonetheless, there are clearly differences between these games, even if it is a gradated matter of degree rather than a stark dividing line.

This distinction is useful for several aspects of historical games studies as different sorts of game engage with history in different ways. Most fundamentally, and as Adam Chapman outlines, a spectrum exists between realist games such as *Assassin's Creed* (which primarily reconstruct history through their audio-visuals and largely fixed narrative) and conceptual games such as *Civilization* (MicroProse, 1991) (which principally explore historical forces and change through the procedural rhetoric of their mechanics) (Chapman 2016: 59–79). Chapman's dichotomy is widely accepted within historical games studies and has exerted substantial influence on the field. It has done much to demonstrate that the design conventions of these different types of game undeniably influence the ways in which they allow players to interact with history. The distinctive characteristics of roleplaying games likewise influence their players' interaction with history and, as this piece will argue, they provide substantial opportunities for the development of historical empathy.

Historical empathy is a relatively well established (Davis et al. 2001; Bryant/Clark 2006; Endacott 2010; 2014; Endacott/Brooks 2018; Hartman et al. 2021), if thoroughly contested and debated (Yilmaz 2007), pedagogical and research concept.[1] Fox's cautious definitions emphasise empathetic approaches as focusing "on people, rather than systems or places" (Fox 2023: 247), present "historicised" empathy as reliant on understanding historical context "to understand and appreciate the worldview of their historical subject" (Ibid.: 247), and seek to codify "historical empathy" (Ibid.: 247–248) as a more detached approach requiring the critical evaluation of the adopted perspective even as it is experienced. These broad definitions are relatively uncontroversial and allow us to understand 'historical empathy' as a means to understand the motivations, behaviours, and ideology of historical characters, rooted in an understanding of their worlds, and achieved through a critical approach. Understanding the cultural and personal mindset of past populations is an important and valid field of study in and of itself,[2] but is also vital in fully comprehending varied subfields of history from economic to environmental to military. Further, historical empathy is intimately connected to the effective scholarly analysis of historical sources (Kohlmeier 2005; Nye et al. 2009), and implicitly or explicitly this empathetic approach forms the core of much historical pedagogy, scholarly writing and research (Fox 2023: 251–264).

The capacity of tabletop roleplaying games and activities to invoke historical empathy is well established. Most studies of this effect revolve around the educational potential of such games and their power has been emphasised as a tool in the classroom at every level (Stanley et al. 2022; Shoenberger 2024; Houghton 2025), including the varied games of the *Reacting to the Past* (Reacting to the Past Consortium 1995-2025) series,[3] but also an absolute plethora of custom-built games (Brynen 2010; Brynen/Milante 2013; Hoy 2018; Mochocki 2021; Konshuh/Klaassen 2022). These games generally require players to take on the role of characters from the appropriate period and act to achieve their goals on the basis of their historical knowledge and research, often encouraging active roleplay as a means of empathetic engagement (Olwell/Stevens 2015: 569–570; Albright 2017: 374; Mochocki 2021: 145; Joseph 2023: 134–135). They typically avoid complex mechanics (Lightcap 2009: 176–177; Morman 2017: 538; Schiffman 2020: 246; Shoenberger 2024: 84–85), and are almost universally embedded in broader curricula through more traditional teaching methods and assignments

---

1 An excellent overview of the main trends in this area is supplied by Fox (2023).
2 As outlined by Belvedresi (2019), Halden/Witcher (2020), Evans (2022: 101–118).
3 See for example: (C. A. Anderson/Dix 2008; Binnington 2015; Joyce et al. 2018; Barber 2023).

(Joseph 2023: 135; Ludolph 2023: 104). The use of analogue roleplaying games in this manner has a long history and, despite some issues, demonstrable success in promoting historical empathy. Outside the classroom, roleplay in reenactment or through casual play has been highlighted as a less formal means to engage with history through empathetic means (Bostal 2019; Mochocki 2021), and a smaller but growing number of scholars have highlighted the potential or actual use of these games as research tools (Cromwell et al. 2021). While the methods and goals vary, in each of these cases, historical empathy is constructed by having players embody their characters, consider the historical context, and exert meaningful control over a historical narrative.

The ability of computer roleplaying games to develop their players' empathy for others has likewise received quite substantial study. These games actively encourage their players to associate themselves with their characters (Burn 2006: 87; Chapman 2020a: 141–143), often through avatar customisation (Mancini/Sibilla 2017), or presentation of a sympathetic protagonist (Jørgensen 2010), alongside multiple dialogue options to allow the player to express themselves through the character (Bessière et al. 2007; Messinger et al. 2008). These attributes have been demonstrated to increase feelings of immersion amongst players and to encourage them to identify and empathise with their characters (McLaughlan/Kirkpatrick 2014; Kline 2016; McLaughlan/Kirkpatrick 2004). Giving players control of characters with different abilities, roles, and backgrounds can fundamentally alter their behaviour and in doing so presents possibilities for the development of empathy with different perspectives (Lucat/Haahr 2015; Domínguez et al. 2016), while granting them control over dialogue and actions, and demonstrating the consequences of these actions gives meaning to player's decisions and can oblige the player to consider the actions and motivations of their character more carefully, often demanding a deeper engagement with the world presented in game (Krzywinska 2008; McKenzie 2018: 54–56). These games also typically place a heavy emphasis on interpersonal relationships, often evolving these over the course of play (Knoll 2018: 142–144; Ryan et al. 2023: 1–2), and in doing so encourage the player to consider the opinions and motivations of characters beyond their own avatar.

However, in spite of their demonstrated ability to invoke empathy amongst their players, the capacity of computer roleplaying games to induce *historical* empathy has received considerably less consideration and study. In fact, roleplaying games have been omitted from many studies of historical computer games: most notably, they are excluded from Chapman's hugely influential theoretical study of history in games (Chapman 2016) and from Wainwrights more practical exploration of the medium (Wainwright 2019). This absence can be explained by a pair of key factors:

Firstly, with a fleeting handful of exceptions such as *Kingdom Come: Deliverance* (Warhorse Studios 2018), *Pentiment* (Obsidian Entertainment 2022), and the *Expeditions* (Logic Artists 2013-2022) series, most roleplaying games have a fantasy or science fiction setting (Rochat 2020: 12–15), and have generally been rejected as non-historical within the scholarship (Alexander 2007: 264). The validity of this rejection has been rebutted by several scholars on the basis that fantasy and history are often blurred in the popular imagination (Young 2010: 163–164; 2018: 1; Chapman 2020b: 91–97; Grufstedt 2023: 37–39), that fantasy settings rely heavily on audience expectations of their period (Salo 2004: 24; Traxel 2008: 125–142), and that fantasy games can have a pronounced impact on their players understanding of historical periods (Houghton 2016).

Secondly, most games of this genre have a medieval or pseudo-medieval setting (Traxel 2008: 127–128), while most studies of historical computer games focus on more modern periods (and the twentieth century in particular). While this is understandable, it does emphasise the importance of considering representations of different historical periods when addressing historical games in general and does not mean that this genre should be ignored.

In any event, digital roleplaying games as history have been discussed comparatively infrequently and generally only within works focusing on the Middle Ages. This is a significant omission as roleplaying games can engage with their worlds and the past in very different ways from those of other genres.

While roleplaying games have been marginalised or ignored within much historical games scholarship, the ability of realist computer games to conduct historical empathy through their immersive qualities has been the subject of a growing body of research and practice. Various scholars have theorised that games such as *Valiant Hearts* (Ubisoft 2014) or *Assassins Creed* (Ubisoft 2007) may invoke historical empathy amongst their players (Rughiniș/Matei 2015; Boltz 2017; S. L. Anderson 2019; Šindelář 2024), emerging from their interactive nature, the agency they grant their players, and the associated sense of immersion which they provide (Rughiniș/Matei 2015: 636; Gilbert 2019: 111; Rodriguez et al. 2021; Patterson et al. 2022; Matei 2023: 13; Richards et al. 2023), and this impact has been demonstrated in both quantitative and qualitative terms (Boltz 2017; Šisler et al. 2022; Stamou et al. 2024; Bowles 2025). Gilbert in particular has demonstrated that playing these games facilitates a deeper understanding of the motivations (Gilbert 2019: 123; Houghton 2023: 6), emotions (Gilbert 2019: 120), humanity (Gilbert 2019: 121–122), and flaws of historical characters (Gilbert 2019: 125–126). This may extend to promoting empathetic understanding for the experiences of subaltern groups or under-represented perspectives across racial (Gilbert 2019: 122–123), class (Gilbert 2019: 124), or religious lines (De Wildt/Aupers 2019: 870–871), or the development of an empathetic understanding of complex debates and conflicts (Boltz 2017: 13; S. L. Anderson 2019: 191; Gilbert 2019: 123; Matei 2023: 13–14; Šindelář 2024: 391). In essence, by experiencing historical events from the perspective of a character living through these events, these scholars argue convincingly that players may develop a deeper and more nuanced understanding of the period.

Most of these studies emphasise the importance of the historical veracity and scholarly authority of their worlds (Rughiniş/Matei 2015: 635; S. L. Anderson 2019: 182–189), demanding that these games follow a fixed narrative with unchangeable and established events (Rughiniş/Matei 2015: 635), and provide extensive historical context (Rughiniş/Matei 2015: 634; Boltz 2017: 14–15). To achieve this, these games lean heavily on the conventions of realist games: building a fixed story and conducting their history through audio-visual fidelity (Stamou et al. 2024: 6) or more stylised approaches (S. L. Anderson 2019: 190) to invoke a sense of the period, often through detailed reproduction of artifacts (Rughiniş/Matei 2015: 636), museum-like descriptions (Rughiniş/Matei 2015: 635; S. L. Anderson 2019: 182–189; Rodriguez et al. 2021) and cinematic techniques (Blečić et al. 2021).

Several scholars have also highlighted how deviation from the conventions of digital games can enhance their ability to induce historical empathy. The use of ordinary protagonists with mundane goals is a potent tool here (Rughiniş/Matei 2015: 635; S. L. Anderson 2019: 191) especially in contrast to the more superhuman characters and spectacular narratives common within games (S. L. Anderson 2019: 191). Slower pacing within these games can allow players more time to reflect and consider the information and experiences they receive from the game world, further establishing its authenticity while also building empathy (Rughiniş/Matei 2015: 637; S. L. Anderson 2019: 182–189; Stamou et al. 2024: 6).

Significantly though, while this scholarship is strong and these realist games can clearly develop players' historical empathy, these studies have almost universally avoided consideration of roleplaying games. This is wholly understandable given the trends within this genre and within the field, but it does represent a missed opportunity – especially as much of the historical empathy gained through realist games is effectively acquired through players roleplaying as their characters to some degree (De Wildt/Aupers 2019: 873). Computer roleplaying games can demonstrably invoke empathy amongst their players and this article will argue that there is little reason why they cannot encourage the development of historical empathy. The chapter will demonstrate this through the discussion and analysis of two case studies in the form of the roleplaying games *Kingdom Come: Deliverance* and *Pentiment*. To this end, the chapter will do three things:

- It will open by briefly describing the two games and justifying their categorisation as roleplaying games.
- It will go on to note and acknowledge the presence of empathetic elements common to realist games within these two roleplaying games.
- It will highlight the ways in which the roleplaying qualities of both games can enhance their ability to invoke historical empathy.

Ultimately, the chapter will argue that roleplaying games and elements are important tools for the development of historical empathy through digital games and should be considered more seriously within the field.

Kingdom Come: Deliverance and *Pentiment*

Kingdom Come and *Pentiment* are relatively recent games and form part of a growing range and diversity of RPGs. As will be outlined below, there are several fundamental differences between the two and they provide substantially different experiences for the player as a result. Nevertheless, they share a handful of core similarities – most notably that they both place heavy emphasis on roleplaying and player agency, and they both play out within a historical rather than fantasy world.

Kingdom Come: Deliverance was released in 2018 by Warhorse Studios – a Czech developer. The player takes control of Henry of Scalitz, the son and apprentice of the village blacksmith whose life is overturned when his village is sacked and his parents killed during the war between Sigismun and Wenceslas in 1403. Henry embarks on a quest to avenge his family and retrieve the sword forged for his lord (Radzig Kobyla), which swiftly escalates to a series of adventures playing a key role in the regional conflict. The game's narrative, appearance and play are at least superficially similar to that of a typical computer roleplaying game – most notably *The Elder Scrolls* (Bethesda 1994-2025) series – with a relatively standard 'rags to riches' storyline crossed with a plot combing personal vengeance and heroic valour, presented through relatively high fidelity graphics, and with gameplay focused on combat, albeit often with other options available (Bostal 2019: 384; Neumann 2019: 71). However, the game is distinguished from the majority of roleplaying games by its historical setting and the vocal commitment to 'historical accuracy' by its developers (Medel 2018: 21–22; Bostal 2019: 381–84; Pfister 2019: 142). The game has been criticised for its depiction of race (Young 2015: 44–45; McCall 2019: 44–45; Testi 2020: 120; Young 2021; Quijano 2023: 197–199), gender (Pfister 2019: 143), and nationalism (Pfister 2020: 60–62), but this focus on historicity has influenced the game in several more constructive ways including its representations of historical sites (Neumann 2019: 71–72), references to historical figures and events, and construction of several mechanics unusual to the genre (Neumann 2019: 72–75; Houghton 2024). Many of these aspects create the potential for the development of historical empathy.

Pentiment is a 2022 game from Obsidian Entertainment – a well-established studio with multiple prior CRPG releases including *Neverwinter Nights 2* (Obsidian Entertainment 2006), *Pillars of Eternity* (Obsidian Entertainment 2015) and *The Outer Worlds* (Obsidian Entertainment 2019). *Pentiment* however is a rather different game from most typical CRPGs. The player is cast as Andreas Maler, an artist from Nuremburg, who has travelled to the monastery at Tassing, a (fictional) town in Bavaria to complete his masterpiece and hence his training. During the game (which spans the period 1518–1543), Andreas becomes embroiled in two murder investigations, before play switches to Magdalene Druckeryn – the daughter of the town printer – who is charged with producing a mural of the town and its history. The game focuses on investigation through various means with very little scope to solve problems through violence, through a story which is quite far removed from that of a typical RPG. Furthermore, *Pentiment* employs a stylised pseudo-medieval illustration for its visuals throughout most of the game – something quite far removed from the increasing emphasis on graphical fidelity in most modern RPGs. The game has been analysed for its engagement with history through its recreation of the world (Meier 2024), connections to medievalist stories, and its use of the construction of history as a game mechanic (Wright 2024). However, the game also centres roleplaying elements and player agency, making it a valuable tool for the development of historical empathy.

These two games are very different in appearance, theme, and gameplay but they both possess elements which set them apart from realist games such as *Assassin's Creed* and *Valiant Hearts*, and they should both be considered roleplaying games. Both *Kingdom Come* and *Pentiment* emphasise the player taking on the role of historical characters, they grant the player a great deal of agency, they provide multiple and varied solutions to the problems faced by the player, they are somewhat non-linear, and they allow the player to make meaningful decisions.

Realist approaches

Kingdom Come and *Pentiment* both employ similar tools for immersion and establishment of historical veracity to those found in the empathetic realist games noted above. Both games employ audio-visual techniques designed to evoke the feel of the period – relying on faithful reproductions and stylised representations respectively. Both employ extensive in game reference material to site their narratives within a claim to historicity, employ lowly protagonists and focus to some extent on mundane activities, and make use of likeable characters and somewhat relatable stories.

Kingdom Come deploys a very similar visual approach to the *Assassin's Creed* series, using reproductions of architecture and material culture to imbue their worlds with apparent historical veracity and invoke historical empathy. In this case creating a detailed model of a section of late medieval Bohemia replete with castles, churches and other buildings based closely on actual sites (Medel 2018: 25; Bostal 2019: 386–388), and detailed recreations of weapons and armour, but also clothing, food items, and tools (Medel 2018: 27; Bostal 2019: 388).

Pentiment's visuals take a very different but still effective approach to evoke historical empathy. The graphics are principally stylised illuminations designed to mimic medieval and early modern designs. These are very far removed from the graphical fidelity of *Kingdom Come: Deliverance* (or the *Assassin's Creed* series), but nonetheless clearly signpost the world's historical period. The effect is heightened by the use of speech bubbles with fonts which correspond to the character's background: educated Andreas uses a carefully inked hand, commoners generally use a scrawled text, the monks and nuns have different fonts reflecting their place of origin, and the town printer and his family use a typeface. Much of the social reality of the game's world is thus portrayed without any recorded lines of speech. In a similar manner to the abstract images and dialogue of *Valiant Hearts*, this approach in *Pentiment* provides an accessible route into the period and a foundation for historical empathy.

These visual approaches are accompanied by the use of audio which is appropriate to the period – or at least which is perceived to be appropriate to the period. The ambient music of both *Kingdom Come* and *Pentiment* employ instruments, effects, and compositions which are typically associated with the later Middle Ages and early modern period with a lot of use of strings, plucked instruments and drums, and *Pentiment* goes so far as to recreate period appropriate compositions. Sound effects associated with the period also appear frequently within both games including Gregorian chanting and natural ambient sounds associated with the Middle Ages such as birdsong or the noises of domestic animal.

Again mirroring approaches common in realist games, both *Kingdom Come* and *Pentiment* make extensive use of a form of historical encyclopaedia to establish their games 'veracity'. *Kingdom Come* includes an in-game 'codex' relating details about individuals, places, and events mentioned or experienced within the game, alongside general information about the world and society of this period. *Pentiment* employs a similar codex, providing details about the key events and background of the game – often delving quite deeply into historical and theological matters – and draws this material close to the gameplay by including rubricated or otherwise highlighted keywords within conversations which immediately link to the relevant codex entries. In *Pentiment*, the codex doesn't merely exist in parallel to the game, it is easily accessible and directly connected (Vandewalle 2024: 110).

The protagonists of both games are relatively lowly, albeit experiencing extraordinary events. Henry in *Kingdom Come* is a commoner while Andreas in *Pentiment* is a member of the emergent middle classes. In both cases, the comparatively humble status of the protagonist is driven home through the plot. Much of the prologue of *Kingdom Come* consists of typical daily duties of Henry as he eats breakfast, help his father collecting supplies and working in the forge, collecting payment for products and engaging in shenanigans with his friends – with a short interlude around learning the basics of swordplay mainly serving to demonstrate Henry's lack of combat ability as an untrained commoner. Likewise, much of the play within *Pentiment* consists of interacting with members of the lower orders of society – often while they work or while sharing a meal with them. In many cases, these interactions, locations, and characters represent subaltern sections of society. *Pentiment* in particular provides a notably broad range of sympathetic and quite detailed characters for the player to interact with including the Jewish printer Benjamin Sommerfeld and his wife Rachel, the Romani traveller Vácslav, and the Ethiopian Brother Sebhat of Sadai. This frequent focus on 'ordinary' characters allows for a development of historical empathy through a very similar approach to that deployed within *Valiant Hearts*: creating characters with whom it is comparatively easy for the player to identify.

Mirroring approaches common within roleplaying games (Jørgensen 2010), the stories and characterisations within both *Kingdom Come* and *Pentiment* are designed to inspire sympathy and empathy with characters. Many of the early quests in *Kingdom Come* revolve around supporting the survivors or mourning the victims of the attack on Skalitz while much of *Pentiment* occurs against the backdrop of the struggles and hopes of the commoners and clergy. Again, this follows a very similar empathetic approach to that used in *Assassin's Creed* and *Valiant Hearts*: it is much easier to develop empathy for fleshed out and likeable characters.

While there are some variations in the approaches of the two games, these aspects closely mirror those of empathetic realist games such as *Assassin's Creed* or *Valiant Hearts*. On this basis alone, they should be considered as potentially useful in developing historical empathy.

Roleplaying opportunities

However, beyond the common features which they share with the empathetic realist games highlighted above, *Kingdom Come* and *Pentiment* incorporate several additional aspects which may promote historical empathy, mainly through their provision of a greater degree of agency to the player by allowing them to make decisions which can alter the outcomes of the narrative. As noted above, the presence of multiple possible outcomes, especially outcomes that are clearly ahistorical, are seen as anathema to the effectiveness of realist games as tools for historical empathy. However, these options in roleplaying games and the agency they provide can allow players to engage with history in a distinct and empathetic manner. In order to facilitate this degree of player agency, these games must often construct more complex or nuanced worlds for the player to explore and must provide the mechanics to allow the player to engage with the world in a variety of ways. The non-linearity of these games can grant additional import to the actions of the player by making their decisions more meaningful, and this can in turn impart an understanding of the motivations behind the actions and decisions of historical characters while also demanding a more reflective engagement with the presented world. This agency also typically extends to character creation and development, supporting a sense of immersion in the period, and often providing distinctive experiences depending on the character played. All of these elements are present in both *Kingdom Come* and *Pentiment* and present potential for the development of historical empathy in a manner distinct from realist games.

The agency which roleplaying games grant to their players generally demands the construction of more detailed worlds and lore than is typical within realist games, and this provides further scope for developing historical empathy (Krzywinska 2008). The core narrative of *Kingdom Come* is effectively linear, but represents only a very small fraction of the game's world with a vast number of side quests, optional activities and characters to interact with. *Pentiment* is more complex, providing multiple routes through the core narrative and including mutually exclusive conversations and activities which demand the game must be played multiple times to experience its entire content. These worlds and their characters are by necessity deeper than those of most of the *Assassin's Creed* games and wider than that of *Valiant Hearts*: RPGs demand a greater level of detail for their players to explore and to support player agency. This degree of complexity enhances their appeal as authoritative depictions of the past and, following the logic applied to realist games, hence augments their ability to inspire historical empathy.

Roleplaying games must also provide mechanical support for these detailed worlds and this can further deepen immersion and promote empathy. *Kingdom Come: Deliverance* provides a range of examples here. Most obviously and basically, and in common with a substantial range of fighting games across genres, the weapons and armour of the game behave very differently depending on their type and purpose. *Kingdom Come* takes this further though with heavy armour physically slowing the player's avatar and several helmets limiting the view presented to the player (Medel 2018: 28–29, Bostal 2019: 389–390). Outside of combat, different items provide different benefits. Darker clothing often boosts the player's ability to move unseen. Softer boots allow them to move more quietly, avoiding the attention of guards, opponents, or hunted animals. More refined or spectacular clothing will change reactions to the player character, making them more persuasive and potentially lowering prices with merchants. The condition of these items also makes a difference in game. Weapons and armour need to be maintained and sometimes repaired in order to remain effective (Medel 2018: 29). Traipsing through the wilderness will dirty clothes, causing the player to be treated with disdain until they wash at a water trough or bath house. In a stark departure from many roleplaying games (Houghton 2024), getting coated with blood through combat will drastically alter the way others react to the player character, making them more intimidating, but also undermining attempts at diplomacy and potentially leading to accusations of murder. All of these elements require the player to engage with the world of the game more carefully, gaining a deeper and more empathetic impression of the period as a result.

Beyond these basic requirements for a more detailed world, the agency granted to players in CRPGs gives greater meaning to their actions with a corresponding impact on their invocation of historical empathy. *Kingdom Come* allows the player to influence the fates of many of the characters they encounter ranging through sparing the lives of enemies, saving the lives of friends or innocents, and developing romantic relationships. The decisions of the player in *Pentiment* will cause the deaths of at least two characters, most of whom are likeable, and whose deaths will drastically change interactions with other characters later in the game – for example, causing the death of Prior Ferenc will lead to a falling out with the abbot making entering the abbey more difficult in the second act, while accusing Lucky

Steinauer the stonemason will sour relations with his wife and daughter leading to markedly different interactions later in the game and potentially closing off some avenues of investigation. Beyond this, the player's less fatal actions throughout the game can have lasting effects including leading to various marriages, allowing a widow to retain her house and land, and influencing a wavering doctor to become an alcoholic or an upstanding member of the community. Seeing these developments unfold as a consequence of the player's actions can immerse them more firmly into the world with a corresponding impact on their development of historical empathy.

Moreover, by placing an emphasis on causes and consequences, these games can provide a deeper explanation for the behaviour of characters (Bessière et al. 2007; Messinger et al. 2008). By allowing players to take these actions, then demonstrating their outcomes (however severe) the game provides a very clear explanation for the motivations and practical limitations of people living within the period. *Kingdom Come* allows the player to berate his social superior Baron Sebastian vom Berg, but this will immediately result in their imprisonment. Likewise, assaulting the inquisitor Bishop Jaroslav of Beneshov will lead to their immediate arrest and execution. *Pentiment* highlights the consequences of openly deviating from Catholic practice through the potential fate of Ursula Gertneryn – a character introduced as an infant at the start of the game and whom the player will interact with throughout the story. Over the course of the game, the player may influence Ursula to adopt pagan or at least heretical practices depending on her interactions with Andreas and several other decisions available to the player. Adopting these heretical views can potentially lead to her execution if she is not also encouraged to keep these practices secret. These harsh outcomes provide clear explanations for the enforcement of the social structure within the worlds of these games and the role of the player in causing these consequences can drive this point home and encourage historical empathy.

Emphasising the impact of the decisions made by the player also demands deeper engagement with the worlds presented within roleplaying games. Players must actively engage with the world around them to understand the likely outcomes of their actions. In many cases, a knowledge of the game world is vital for gaining the best endings or outcomes for individual characters and this can motivate players to engage with the world more deeply and reflect on their actions to a much greater extent (McKenzie 2018: 54–56). The various murder mysteries in *Pentiment* actively demand that the player engage with the game world and its historical elements in order to draw their conclusions: they must select which characters to investigate and which leads to follow, and in doing so they are obliged to become familiar with various aspects of the game's early modern setting. The various lines of enquiry can lead the player to investigating Roman ruins, medieval and classical literature, various aspects of theology, and countless elements of minutiae around local society and politics. To get the most information, players must consult with a wide range of individuals from across societal roles and often originating from outside the region. The narrative approaches designed to make the player care about characters and identify with their own avatars, can drive players to consider and reflect on these details carefully and form an important element in developing historical empathy.

The extension of player agency to the customisation of avatars in these games also plays a significant role in promoting identification with these characters and hence in developing historical empathy. Both *Kingdom Come* and *Pentiment* require the player to use specific avatars with relatively limited capacity to customise appearance, but both nevertheless allow the creation of very different characters. *Kingdom Come* follows the typical RPG model of 'levelling up', granting the player experience points to earn levels which allow them to select 'perks' and other abilities to customise their character. Further, the various weapons, clothing, and other equipment can substantially alter Henry's appearance and abilities when equipped. *Pentiment* instead places an emphasis on Andreas' background experience prior to the events of the game and during interludes. In both cases, there is plenty of scope for the player to craft a character matching their interests, and potentially inserting elements of themselves into their characters. Mancini and Sibilla have demonstrated a strong correlation between this sort of customisation and the degree of immersion felt by the player (Mancini/Sibilla 2017), and this customisation can allow players to develop a degree of empathy with their characters (McLaughlan/Kirkpatrick 2002; Kline 2016; McLaughlan/Kirkpatrick 2014).

Moreover, the variety of different pathways available for character customisation can create very different experiences depending on the background of the player's avatar within both *Kingdom Come* and *Pentiment* and, through the course of multiple playthroughs, this can develop a broad historical empathy through the consideration of multiple viewpoints within the game. This is visible to some degree through the levelling up system of *Kingdom Come* which allows the player to specialise in different forms of combat, but also in stealth and diplomacy creating potentially different experiences throughout the game as the player is increasingly drawn to solutions to challenges which match their skillsets: burly players attacking enemies head on, sneaky players evading or poisoning their foes, and diplomatic players talking their way out of trouble, thereby receiving correspondingly different results. A number of acquirable 'perks' can further alter the player's experience: 'Highborn' will grant better results when speaking to nobles while 'lowborn' will support interactions with commoners. On top of this, playing through the game while remaining illiterate is entirely possible and gives a distinct (if somewhat limited) perspective of the world (Medel 2018: 29). The impact of character customisation is more readily apparent in *Pentiment* where the player must choose several facets of Andreas' background including the locations of their career, their interests, their Master's degree, and their favourite academic subjects. Each of these decisions impacts the game, determining the languages Andreas can understand, their knowledge of pivotal subjects, and various other factors around their persuasiveness in different circumstances. These different backgrounds can hugely influence the information uncovered by Andreas during his investigations – whether through observation, research, or persuasive investigation – and often impact the actions available to him. In both games, different playthroughs may result in drastically different experiences depending on how the character has been customised, providing opportunities for the exploration of different viewpoints.

Both *Kingdom Come* and *Pentiment* develop these distinctive experiences by granting the player control of a second protagonist of a different gender and with a very different experience during play. During the *A Woman's Lot* DLC for *Kingdom Come*, the player embodies Theresa, the Miller's daughter in Skalitz throughout the events during the prologue of the main game: undertaking daily activities before surviving the attack on the town. This provides an immediate mirror to the activities undertaken by Henry during the same period and serves to present some stark differences between male and female experiences. In contrast with Henry's being woken, fed, and sent to work by his mother, Theresa is awake and doing chores before her father and must wake her brother Samuel and chide the labourer Zbyshek to get to work before continuing a range of chores. Theresa's encounters with the various other members of the village are rather different from Henry's with curter orders from the men and more opportunity to gossip with women, while her abilities are focused on diplomacy, stealth and healing, with very limited combat abilities and being completely unable to ride a horse or wear armour. In a similar manner, the change of protagonist from Andreas to Magdalene in the final act of *Pentiment* presents several core differences in the experiences and attitudes of the two characters. Magdalene faces immediate (and unfounded) questions about her ability to conduct her work on the basis of her gender and a subplot around her potential marriage. Her background is constructed in much the same manner as Andreas, but the available choices are reflective of Magdalene's status as the daughter of a printer in a small town, with a focus on the work that she has done (including 'Domestics' and 'Bookkeeping'), her reputation in the town (such as reputations for slinging 'Barbs', or being a 'Flirt' or a 'Snoop') and reading interests (rather than formal education). The player's experience as Magdalene is distinct from that when playing Andreas, with different interactions with the various characters throughout the setting. In both cases, the distinction between the experiences of male and female characters is rooted to some extent in the general constraints and societal expectations applied to women's actions during the period of the games, although this may be exaggerated by the expectations of the developers and the male-centric conventions of the genre. In any event, the games emphasise the existence of a diversity of experiences within their settings.

The effect in each of these cases is that players' experiences of the game and its world and narrative are starkly different depending on their character's background and choices, and this can emphasise the differences and similarities between different elements of society. Several scholars have demonstrated that taking on a different role or character within a roleplaying game can substantially influence player behaviour as they take on attitudes and approaches which match those associated with their chosen character (Yee et al. 2009; Lucat/Haahr 2015; Domínguez et al. 2016). This can foster distinct roleplay approaches depending on the different character played which, combined with these very different experiences and the historicised foundations of these games, can provoke a substantial degree of historical empathy. In essence, these distinctions between backgrounds in roleplaying games combine the strength of empathy building provided through character customisation with the potency of providing diverse perspectives in other empathetic games outlined above.

Ultimately then, while there are several key overlaps between the ways in which roleplaying games such as *Kingdom Come* and *Pentiment* and realist games such as *Assassin's Creed* and *Valiant Hearts* provide avenues for the development of historical empathy, these roleplaying games also possess a range of additional qualities which can encourage the emergence of this empathy in distinct and potentially more potent ways.

Conclusion

Computer roleplaying games present a substantial opportunity for the development of historical empathy. This can certainly be accomplished through the well-established approaches of realist games which employ methods drawn from museums and other forms of public history, enhanced through the interactive and immersive qualities of digital games. However, there is also potential to use computer roleplaying games to encourage historical empathy through other means: the depth of worlds, character customisation, and, above all, the agency granted to players within the roleplaying genre. *Kingdom Come* and *Pentiment* are very different games, but both highlight this potential.

There are of course issues with the use of both *Kingdom Come* and *Pentiment* for these empathetic purposes. The meaningful decisions so central to both games presents an ambiguity of history by allowing the player to change historical events which runs counter to the common assertions made around the necessity of a fixed narrative within empathetic realist games exemplified by Rughiniş and Matei (2015) above. Beyond this, both games deviate from history to some extent: *Kingdom Come* changing the chronology of its events to fit its narrative, and *Pentiment* creating an entirely fictitious town as the centre of the game. There are also more specific issues relating to the two games. For example, there are several concerns around the representation of subaltern groups in *Kingdom Come*, while the multiple unknowables within the plot of *Pentiment* can prove a frustration which undermines historical empathy amongst some players.

There are further issues within roleplaying games more broadly. As underlined above, *Kingdom Come* and *Pentiment* are part of a small minority of roleplaying games which employ a historical setting with the vast majority of this genre using fantasy or science fiction worlds, and these obviously ahistorical settings have been held to undermine historical empathy. The focus on violence as principal activity and the solution to most problems within these games can likewise limit their utility in this regard, while the well documented issues around representations of race, gender and sexuality within the roleplaying genre can create further difficulties. The degree of character customisation available in many of these games can encourage self-insertion which may undermine attempts to identify with and understand figures from different eras, and the presence of meaningful decisions with multiple outcomes can encourage players to concentrate on the results of actions rather than their motivations, potentially leading to inconsistent and arbitrary behaviour to achieve the desired outcome. Perhaps most blatantly, there is a marked tendency within these games to focus on superhuman characters undertaking quests to save the world (Marino Carvalho 2016: 113–114), and this can create a vision of history as driven by the actions of a handful of great individuals which can undermine attempts at developing empathy.

However, this genre provides an important alternative approach to history and historical empathy within digital games which should not be ignored, especially as many of these issues can be mitigated or resolved. Most basically, a fictional historical narrative with multiple possible outcomes is not automatically less believable or immersive than a fictional narrative with a set outcome – such as that used within *Valiant Hearts* – indeed, there is indisputably value to games which break from known historical narratives through counterfactual history (McCall 2012: 17; Chapman 2016: 231–257; Metzger/Paxton 2016: 547–548; Apperley 2018: 9–12), and the presence of competing and conflicting accounts of historical events can force the player to develop their own understanding of the world based on deep reflection on the rival accounts they experience (DiPietro 2014: 209–211). In a similar vein, despite their marginalisation, fantasy games can provide useful historical empathy as they are deeply associated with historical periods – especially the Middle Ages (Tolmie 2006: 149–150; Young 2010: 163–164; 2018: 1) – and are often accepted by players as presenting a valid depiction of certain aspects of history (Chapman 2020b: 92–97; Grufstedt 2023: 37–39), not to mention their utility in engaging players with historical myths and mythologies (Vandewalle 2024). The remaining difficulties can largely be addressed through selection of games and critical play (Flanagan 2013; McCall 2016).

This empathetic potential of roleplay can be extended well beyond the roleplaying genre. As highlighted above, the core empathetic approach of realist games is in essence a form of roleplay. Indeed, many realist historical games make use of these empathetic elements, albeit typically to a much lesser extent than those mentioned in this article. Likewise, a growing number of conceptual games employ roleplaying elements to varying extents and hence provide greater potential for their players to engage in historical empathy. *Crusader Kings III* (Paradox Development Studio 2020) is an important example of this through its detailed characterisation of historical and pseudo-historical figures supported with mechanical consequences for acting out of character and combined with a vast and detailed set of conceptual mechanics.

This study has highlighted a consequence of siloing scholarly approaches to history through games. The study of historical empathy through digital games is an important and emergent field and its connection to museum and heritage studies has proven very fruitful. However, this scholarship has largely ignored the body of work addressing historical empathy through analogue roleplay and this has rather limited the scope of the field. More broadly, there is a clear case that empathetic approaches can be profitably connected to the realist and conceptual model proposed by Chapman (2016) and the varied studies which have been built on this basis. Extending these links promise valuable perspectives in each of these connected arenas.

Ultimately, historical empathy is an important component of historical understanding and digital games provide several opportunities to develop this. Bierstedt has rightly noted that communicating historical mentalities is a substantial challenge in games in general (Bierstedt 2023: 141–142). Nevertheless, this is a worthwhile activity, and computer roleplaying games provide a viable opportunity to deliver a form of historical empathy.

References

Albright, C. L. (2017): "Harnessing Students' Competitive Spirit: Using Reacting to the Past to Structure the Introductory Greek Culture Class." The Classical Journal 112(3), pp. 364–379.

Alexander, M. (2007): Medievalism: The Middle Ages in Modern England. Yale University Press.

Anderson, C. A./ Dix, T. K. (2008): "'Reacting to the Past' and the Classics Curriculum: Rome in 44 BCE." The Classical Journal 103(4), pp. 449–455.

Anderson, S. L. (2019): "The Interactive Museum: Video Games as History Lessons through Lore and Affective Design." E-Learning and Digital Media 16(3), pp. 177–295.

Apperley, T. (2018): "Counterfactual Communities: Strategy Games, Paratexts and the Player's Experience of History." Open Library of Humanities 4(1), pp. 1–17.

Arjoranta, J. (2011): "Defining Role-Playing Games as Language-Games." International Journal of Role-Playing 2, pp. 3–17.

Barber, C. (2023): "Reflecting on Reacting: Incorporating Reacting to the Past Games in the High School Classroom." Teaching History: A Journal of Methods 48(1), pp. 76–83.

Belvedresi, R. E. (2019): "Empathy and Historical Understanding: A Reappraisal of 'Empathic Unsettlement.'" In: R. G. Aguilar (ed.), Empathy: Emotional, Ethical and Epistemological Narratives. Leiden: Brill, pp. 162–177.

Bessière, K./Fleming Seay, A./Kiesler, S. (2007): "The Ideal Elf: Identity Exploration in World of Warcraft." CyberPsychology & Behavior 10(4), pp. 530–535.
Bethesda (1994-2025): The Elder Scrolls series [Video Game]. Bethesda Softworks.
Bierstedt, A. (2023): "Making Friendships, Breaking Friendships: Exploring Viking-Age Social Roles through Player Strategy in A Feast for Odin." In: R. Houghton (ed.), Playing the Middle Ages: Pitfalls and Potential in Modern Games. New York: Bloomsbury Academic, pp. 131-148.
Binnington, I. (2015): "Teaching with Reacting to the Past: Bringing Role-Immersion Play into the College Classroom." The Journal of the Gilded Age and Progressive Era 14(4), pp. 589–592.
Blečić, I./ Cuccu, S./ Fanni, F. A. et al. (2021): "First-Person Cinematographic Videogames: Game Model, Authoring Environment, and Potential for Creating Affection for Places." Journal on Computing and Cultural Heritage 14(2), pp. 1–29.
Boltz, L. O. (2017): "'Like Hearing from Them in the Past': The Cognitive-Affective Model of Historical Empathy in Videogame Play." International Journal of Gaming and Computer-Mediated Simulations 9(4), pp. 1–18.
Bostal, M. (2019): "Medieval Video Games as Reenactment of the Past: A Look at Kingdom Come: Deliverance and Its Historical Claim." Congreso Asociación de Historia Contemporánea - Actas 14, pp. 380-394.
Bowles, H. (2025): "Evaluating Bubbles: An Evaluation of a New Immersive Indirect E-Contact Learning Platform." MRes, Kent.
Bryant, D./Clark, P. (2006): "Historical Empathy and 'Canada: A People's History.'" Canadian Journal of Education / Revue Canadienne de l'éducation 29(4), no. 1039.
Brynen, R. (2010): "(Ending) Civil War in the Classroom: A Peacebuilding Simulation." PS: Political Science & Politics 43(1), pp. 145–149.
Brynen, R./Milante, G. (2013): "Peacebuilding with Games and Simulations." Simulation & Gaming 44(1), pp. 27–35.
Burn, A. (2006): "Playing Roles." In D. Carr (ed.), Computer Games: Text, Narrative, and Play. Cambridge: Polity.
Chapman, A. (2016): Digital Games as History: How Videogames Represent the Past and Offer Access to Historical Practice. New York: Routledge.
Chapman, A. (2020a): "Playing Against the Past? Representing the Play Element of Historical Cultures in Video Games." In: A. von Lünen/Lewis, K. J./ Litherland, B./ Cullum, P. H. (eds.), Historia Ludens: The Playing Historian. New York: Routledge.
Chapman, A. (2020b): "Playing the Historical Fantastic: Zombies, Mecha-Nazis and Making Meaning about the Past through Metaphor." In: Hammond, P./ Pötzsch, H. (eds.), War Games: Memory, Militarism and the Subject of Play. New York: Bloomsbury Academic.

Cromwell, J./Dar, A./ Hayes, E./ Ochała, G./ Scheerlinck. E. (2021): "'Dice on the Nile': Roleplaying History." Paper presented at Manchester Game Studies Network, Manchester. October 5.
Davis, O. L./Yeager, E. A./ Foster, S. J. (eds.) (2001): Historical Empathy and Perspective Taking in the Social Studies. Lanham: Rowman & Littlefield.
De Wildt, L./Aupers, S. (2019): "Playing the Other: Role-Playing Religion in Videogames." European Journal of Cultural Studies 22(5–6), pp. 867–884.
DiPietro, M. (2014): "Author, Text, and Medievalism in The Elder Scrolls." In: D. T. Kline (ed.), Digital Gaming Re-Imagines the Middle Ages. New York: Routledge.
Domínguez, I. X./Cardona-Rivera, R . E./ Vance, J. K./Roberts, D. L. (2016): "The Mimesis Effect: The Effect of Roles on Player Choice in Interactive Narrative Role-Playing Games." CHI '16: Proceedings of the 2016 CHI Conference on Human Factors in Computing Systems (New York, NY), pp. 3438–3449.
Endacott, J. L. (2010): "Reconsidering Affective Engagement in Historical Empathy." Theory & Research in Social Education 38(1), pp. 6–47.
Endacott, J. L. (2014): "Negotiating the Process of Historical Empathy." Theory & Research in Social Education 42(1), pp. 4–34.
Endacott, J. L./Brooks, S. (2018): "Historical Empathy: Perspectives and Responding to the Past." In: S. A. Metzger/L. M. Harris (eds.), The Wiley International Handbook of History Teaching and Learning. Hoboken: Wiley. pp. 203-225.
Evans, T. (2022): Family History, Historical Consciousness and Citizenship: A New Social History. New York: Bloomsbury Academic.
Flanagan, M. (2013): Critical Play: Radical Game Design. Cambridge: MIT Press.
Fox, S. (2023): "Archival Intimacies: Empathy and Historical Practice in 2023." Transactions of the Royal Historical Society 1, pp. 241–265.
Gilbert, L. (2019): "'Assassin's Creed Reminds Us That History Is Human Experience': Students' Senses of Empathy While Playing a Narrative Video Game." Theory & Research in Social Education 47(1), pp. 108–137.
Grufstedt, Y. (2023): "Unbending Medievalisms – Finding Counterfactual History in Sandbox Games Set in the Middle Ages." In: R. Houghton (ed.), Playing the Middle Ages: Pitfalls and Potential in Modern Games. New York: Bloomsbury Academic, pp. 29–50.
Halden, D. van/Witcher, R. (2020): "Walking in Someone Else's Shoes: Archaeology, Empathy and Fiction." In: D. van Halden/R. Witcher (eds.), Researching the Archaeological Past through Imagined Narratives: A Necessary Fiction. New York: Routledge.
Hartman, A./Tulloch, R./Young, H. (2021): "Video Games as Public History: Archives, Empathy and Affinity." Game Studies 21(4), n. p.
Hitchens, M./Drachen, A. (2008): "The Many Faces of Role-Playing Games." The International Journal of Role-Playing 1, pp. 3–21.

Houghton, R. (2016): "Where Did You Learn That? The Self-Perceived Educational Impact of Historical Computer Games on Undergraduates." Gamevironments 5, pp. 8–45.

Houghton, R. (2023): "Awesome, but Impractical? Deeper Engagement with the Middle Ages through Commercial Digital Games." Open Library of Humanities 9(2).

Houghton, R. (2024): "A Violent Medium for a Violent Era: Brutal Medievalist Combat in Dragon Age: Origins and Kingdom Come: Deliverance." Studies in Medievalism 33, pp. 119-144.

Houghton, R. (2026): "Getting into Character: Educational Mechanics and Roleplay in the Medieval History Classroom." In: T. Gobbitt (ed.), Tabletop Gaming the Medieval World: Reception, Education and Adaptation. Leiden: Brill.

Hoy, B. (2018): "Teaching History With Custom-Built Board Games." Simulation & Gaming 49(2), pp. 115–133.

Jørgensen, K. (2010): "Game Characters as Narrative Devices. A Comparative Analysis of Dragon Age: Origins and Mass Effect 2." Eludamos: Journal for Computer Game Culture 4(2), pp. 315–331.

Joseph, P. (2023): "Is Including Games in a College Class Beneficial?" Teaching History: A Journal of Methods 48(1), pp. 134–135.

Joyce, K. E./Lamey, A./Martin, N. (2018): "Teaching Philosophy through a Role-Immersion Game: Reacting to the Past." Teaching Philosophy 41(2), pp. 175–198.

Kline, D. T. (2016): "Participatory Medievalism, Role-Playing, and Digital Gaming." In: L. D'Arcens (ed.), The Cambridge Companion to Medievalism. Cambridge University Press, pp. 75–88.

Knoll, T. (2018): "'Instant Karma' - Moral Decision Making Systems in Digital Games." Religions 9(4), pp. 126–160.

Kohlmeier, J. (2005): "The Impact of Having 9th Graders 'Do History.'" The History Teacher 38(4), pp. 499–524.

Konshuh, C./Klaassen, F. (2022): "The Renaissance Marriage Game: A Simulation for Large Classes." In: R. Houghton (ed.), Teaching the Middle Ages through Modern Games: Using, Modding and Creating Games for Education and Impact. Berlin: De Gruyter, pp. 155–171.

Krzywinska, T. (2008): "World Creation and Lore: World of Warcraft as Rich Text." In: H. Corneliussen/J. W. Rettberg (eds.), Digital Culture, Play, and Identity: A World of Warcraft Reader. Cambridge: MIT Press, pp. 123–141.

Lightcap, T. (2009): "Creating Political Order: Maintaining Student Engagement through Reacting to the Past." Political Science & Politics 42(1), pp. 175–179.

Logic Artists (2013-2022): Expeditions series [Video Game]. BitComposer.

Lucat, B./Haahr, M. (2015): "What Makes a Successful Emergent Narrative: The Case of Crusader Kings II." In: H. Schoenau-Fog/L. E. Bruni/S. Louchart/S. Baceviciute (eds.), Interactive Storytelling. Cham: Springer International Publishing, pp. 259–266.

Ludolph, P. (2023): "Measuring Critical Thinking in Reacting to the Past." Teaching History: A Journal of Methods 48(1), pp. 91–105.
Mancini, T./Sibilla, F. (2017): "Offline Personality and Avatar Customisation. Discrepancy Profiles and Avatar Identification in a Sample of MMORPG Players." Computers in Human Behavior 69, pp. 275–283.
Marino Carvalho, V. (2016): "History and Human Agency in The Witcher 3: Wild Hunt." Gamevironments 5, pp. 104–131.
Matei, Ș. (2023): "The Technological Mediation of Collective Memory Through Historical Video Games." Games and Culture 20(4), pp. 477–488.
McCall, J. (2012): "Navigating the Problem Space: The Medium of Simulation Games in the Teaching of History." History Teacher 1, pp. 9–28.
McCall, J. (2016): "Teaching History with Digital Historical Games: An Introduction to the Field and Best Practices." Simulation & Gaming 47(4), pp. 517–542.
McCall, J. (2019): "Playing with the Past: History and Video Games (and Why It Might Matter)." Journal of Geek Studies 6(1), pp. 29–48.
McKenzie, A. (2018): "A Patchwork World: Medieval History and World-Building in Dragon Age: Inquisition." The Year's Work in Medievalism 33, pp. 49–61.
McLaughlan, R. G./Kirkpatrick, D. (2004): "Online Roleplay: Design for Active Learning." European Journal of Engineering Education 29(4), pp. 477–490.
McLaughlan, R. G./Kirkpatrick, D. (2014): "Peer Learning Using Computer Supported Roleplay Simulations." In: D. Boud/R. Cohen/J. Sampson (eds.), Peer Learning in Higher Education: Learning from and with Each Other. Hoboken: Taylor and Francis, pp. 141–155.
Medel, I. (2018): "Kingdom Come: Deliverance y La Representación de La Baja Edad Media." E-Tramas 1, pp. 17–33.
Meier, J. (2024): "Nichts mehr zu verlieren: Die Darstellung von Rechtsbewusstsein und Gemeindekonflikten in Pentiment." Mittelalter. Interdisziplinäre Forschung und Rezeptionsgeschichte 4, pp. 85–99.
Messinger, P. R./Ge, X./Stroulia, E./Lyons, K./Smirnov, K./Bone, M. (2008): "On the Relationship between My Avatar and Myself." Journal for Virtual Worlds Research 1(2).
Metzger, S. A./Paxton, R. J. (2016): "Gaming History: A Framework for What Video Games Teach About the Past." Theory & Research in Social Education 44(4), pp. 532–564.
MicroProse (1991): Civilization [Video Game]. MicroProse.
Mochocki, M. (2021): Role-Play as a Heritage Practice: Historical Larp, Tabletop RPG and Reenactment. New York: Routledge.
Morman, T. A. (2017): "Negotiation Simulation Games for Any History Class." The History Teacher 50(4), pp. 535–549.
Neumann, M. (2019): "Representation of Medieval Realia in PC Game: Kingdom Come: Deliverance." Czech-Polish Historical and Pedagogical Journal 11(2), pp. 69–76.

Nye, A./ Hughes-Warrington, M./Roe, J. et al. (2009): "Historical Thinking in Higher Education: Staff and Student Perceptions of the Nature of Historical Thinking." History Australia 6(3), pp. 73.1–73.16.

Obsidian Entertainment (2006): Neverwinter Nights 2 [Video Game]. Atari Interactive.

Obsidian Entertainment (2015): Pillars of Eternity [Video Game]. Paradox Interactive.

Obsidian Entertainment (2019): The Outer Worlds [Video Game]. Private Division.

Obsidian Entertainment (2022): Pentiment [Video Game]. Xbox Game Studios.

Paradox Development Studio (2020): Crusader Kings III [Video Game]. Paradox Interactive.

Patterson, T./Han, I./Esposito, L. (2022): "Virtual Reality for the Promotion of Historical Empathy: A Mixed-Methods Analysis." Theory & Research in Social Education 50(4), pp. 553–580.

Pfister, E. (2019): "Kingdom Come Deliverance: A Bohemian Forest Simulator." Gamevironments 11, pp. 142–148.

Pfister, E. (2020): "Why History in Digital Games Matters: Historical Authenticity as a Language for Ideological Myths." In: M. Lorber/F. Zimmermann (eds.), History in Games: Contingencies of an Authentic Past. Bielefeld: transcript, pp. 47–72.

Quijano, J. (2023): "Virtually (de)Colonized: Racial Identity and Colonialism in the Middle Ages and as Depicted in Kingdom Come: Deliverance, A Plague Tale: Innocence, The Elder Scrolls, and Black Desert Online." In: R. Houghton (ed.), Playing the Middle Ages: Pitfalls and Potential in Modern Games. New York: Bloomsbury Academic, pp. 195–211.

Reacting to the Past Consortium (1995-2025): Reacting to the Past series [Tabletop Game]. Reacting Consortium Press.

Richards, D./ Lupack, S./ Bilgin, A. A. B./Bronwen, N./Porte, M. (2023): "Learning with the Heart or with the Mind: Using Virtual Reality to Bring Historical Experiences to Life and Arouse Empathy." Behaviour & Information Technology 42(1), pp. 1–24.

Rochat, Y. (2020): "A Quantitative Study of Historical Video Games (1981–2015)." In: A. von Lünen/K. J. Lewis/B. Litherland/P. H. Cullum (eds.), Historia Ludens: The Playing Historian. New York: Routledge, pp. 3–19.

Rodriguez, A./Chuchiak, J. F./Erickson, H. B./ Ballicu, P. V. (2021): "The Great 'Auto de Fe' at Santiago de Los Caballeros, or How to Achieve Historical Empathy with Cultural Heritage through Virtual Reality." The International Archives of the Photogrammetry, Remote Sensing and Spatial Information Sciences 46, pp. 633–640.

Rughiniș, R./ Matei, S. (2015): "Play to Remember: The Rhetoric of Time in Memorial Video Games." In: M. Kurosu (ed.), Human-Computer Interaction: Interaction Technologies. Cham: Springer International Publishing, pp. 628–639.

Olwell, R./Stevens, A. (2015): "'I Had to Double Check My Thoughts': How the Reacting to the Past Methodology Impacts First-Year College Student Engagement, Retention, and Historical Thinking." The History Teacher 48(3), pp. 561–572.

Ryan, M./McEwan, M./ Formosa, P./ Messer, J./Howarth, S. (2023): "The Effect of Morality Meters on Ethical Decision Making in Video Games: A Quantitative Study." Computers in Human Behavior 142, art. 107623.

Salo, D. (2004): "Heroism and Alienation through Language in The Lord of the Rings." In: M. W. Driver/S. Ray (eds.), The Medieval Hero on Screen: Representations from Beowulf to Buffy. Jefferson: McFarland, pp. 23–37.

Schiffman, J. R. (2020): "Early Design Challenges in Developing a Reacting to the Past Game." The History Teacher 53(2), pp. 239–253.

Shoenberger, M. (2024): "Tertiary Learning Usage of Tabletop Roleplaying Games: Affordances and Challenges." In T. Bowell/N. Pepperell/A. Richardson/M-T. Corino (eds.), Revitalising Higher Education: Insights from Te Puna Aurei LearnFest 2022. Cardiff University Press, pp. 80–86.

Šindelář, J. (2024): "Playing through to Europe? Depiction and Reception of the First World War in the Videogame Valiant Hearts." Journal of Contemporary European Studies 32 (2), pp. 386–399.

Šisler, V./ Pötzsch, H./Hannemann, T./Cuhra, J./Pinkas, J. (2022): "History, Heritage, and Memory in Video Games: Approaching the Past in Svoboda 1945: Liberation and Train to Sachsenhausen." Games and Culture 17(6), pp. 901–914.

Stamou, S./Apostolakis, K. C./Ntoa, S./Margetis, G./Stephanidis, C. (2024): "Museum-Inspired Video Games as a Symbolic Transitional Justice Policy: Overview, Concepts, and Research Directions." Games: Research and Practice 2(2), pp. 1–17.

Stanley, E./Sweeten, D./Schmidt, M. (2022): "Embracing the Fiasco!: Roleplaying Games, Pedagogy and Student Success." Dialogue: The Interdisciplinary Journal of Popular Culture and Pedagogy 9(4), pp. 18–39.

Testi, D. (2020): "'Perros Paganos' La Historia de Los Merodeadores Cumanos de 'Kingdom Come Deliverance.'" Humanidades Digitales y Videojuegos 9, pp. 119–136.

Tolmie, J. (2006): "Medievalism and the Fantasy Heroine." Journal of Gender Studies 15(2), pp. 145–158.

Traxel, O. M. (2008): "Medieval and Pseudo-Medieval Elements in Computer Role-Playing Games: Use and Interactivity." Studies in Medievalism 16, pp. 125–142.

Ubisoft (2007-2025): Assassin's Creed series [Video Game]. Ubisoft.

Ubisoft (2007): Assassin's Creed [Video Game]. Ubisoft.

Ubisoft (2018): Assassin's Creed: Odyssey [Video Game]. Ubisoft.

Ubisoft (2020): Assassin's Creed: Valhalla [Video Game]. Ubisoft.

Ubisoft (2025): Assassin's Creed: Shadows [Video Game]. Ubisoft.

Ubisoft (2014-2023): Valiant Hearts series [Video Game]. Ubisoft.

Ubisoft (2014): Valiant Hearts: The Great War [Video Game]. Ubisoft.
Vandewalle, A. (2024): "Video Games as Mythology Museums? Mythographical Story Collections in Games." International Journal of the Classical Tradition 31(1), pp. 90–112.
Wainwright, A. M. (2019): Virtual History: How Videogames Portray the Past. New York: Routledge.
Warhorse Studios (2018): Kingdom Come: Deliverance [Video Game]. Deep Silver.
Wright, E. (2024): "'Layers of History': History as Construction/ Constructing History in Pentiment." ROMchip: A Journal of Game Histories 6(1).
Yee, N./Bailenson, J. N./Ducheneaut, N. (2009): "The Proteus Effect: Implications of Transformed Digital Self-Representation on Online and Offline Behavior." Communication Research 36(2): pp. 285–312.
Yilmaz, K. (2007): "Historical Empathy and Its Implications for Classroom Practices in Schools." The History Teacher 40(3), pp. 331–337.
Young, H. V. (2010): "Approaches to Medievalism: A Consideration of Taxonomy and Methodology through Fantasy Fiction." Parergon 27(1), pp. 163–179.
Young, H. V. (2015): "Whiteness and Time: The Once, Present and Future Race." Studies in Medievalism 24, pp. 39–50.
Young, H. V. (2018): Race and Popular Fantasy Literature: Habits of Whiteness. New York: Routledge.
Young, H. V. (2021): "Race and Historical Authenticity Kingdom Come: Deliverance." In: K. Alvestad/R. Houghton (ed.), The Middle Ages in Modern Culture: History and Authenticity in Contemporary Medievalism. New York: Bloomsbury, pp. 28–39.
Zagal, J. P./Deterding, S. (2018): "Definitions of 'Role-Playing Games.'" In: J. P. Zagal/S. Deterding (eds.), Role-Playing Game Studies: Transmedia Foundations. New York: Routledge, pp. 19–51.

Revisioning, Reframing, Coding the Past

The Truth(iness) is a Lie
Historical Re-visions of the Cold War through *Call of Duty* Paratexts

victoria l. braegger and Samantha Blackmon

Abstract

A Chia Pets advertisement. A news report on median home prices in 1991. Bill Clinton playing the saxophone. The title sequence for the television show Frank's Woods? Months before *Call of Duty: Black Ops 6*'s release date, Activision began building their version of the Cold War and the historical period through paratexts. In this article, we examine the first three trailers for *Black Ops 6* for how they re-vision history by using truthiness, or the feeling that something is true rather than truths built on objective facts (à la Stephen Colbert), as a narrative tool. The first trailer, wake up (rushmore), sets the historical period through cultural foundations. The second release trailer, Open Your Eyes, blurs the lines between reality and gameworlds by blending archival footage with game narrative. Clips of world leaders, 1990s popular culture, and false clips (such as the opening for the above-mentioned television show Frank's Woods) tell the viewer, 'The truth lies. You must find the truth'. Lines are further blurred in the third trailer, The Truth Lies - Live Action Reveal Trailer, where real clips of helicopters and targeting screens are meshed with deepfake-level computer-generated versions of world leaders; Colin Powell mocks the viewer: "You wanted peace, so we hid the war" Despite its name, the live action trailer is anything but.

Games, like *Call of Duty*, that utilise historical events offer the opportunity to engage with and build interest in history, especially in populations that have no prior interest (and little prior knowledge) of the historical period. Simultaneously, this revisionist history becomes the only history that some of the aforementioned population knows. The blending of actual historical references and deepfakes (even those that are unintentional) ultimately reinforces the truthiness of this revised history. In this article, we use depictions of the Cold War as presented in *Call of Duty: Black Ops 6*'s paratexts as a lens through which to explore and examine the problems and potentialities of using actual historical events in the processes of worldbuilding in game IPs like *Call of Duty*.

Keywords

Call of Duty, video game paratexts, video game trailers, truthiness, historical revision

1. Who put the truth in truthiness?

> The human race is becoming a disgrace.
> The rich get richer.
> The poor are getting poorer.
> Fascist, chauvinistic government fools.
> People, Moslems, Christians and Hindus.
> Are in a time zone just searching for the truth.
> Who are you to think you're a superior race?
>
> "World Destruction," Afrika Bambaataa

> And on this show, on this show your voice will be heard in the form of my voice. 'Cause you're looking at a straight-shooter, America. I tell it like it is. I calls 'em like I sees 'em. I will speak to you in plain simple English.
> And that brings us to tonight's word: Truthiness.
> Now I'm sure some of the Word Police, the wordanistas over at Webster's, are gonna say, "Hey, that's not a word". Well, anybody who knows me knows that I'm no fan of dictionaries or reference books. They're elitist. Constantly telling us what is or isn't true, or what did or didn't happen. Who's Britannica to tell me the Panama Canal was finished in 1914? If I wanna say it happened in 1941, that's my right.
> I don't trust books. They're all fact, no heart.
>
> Stephen Colbert (as cited in Zimmer 2005)

In the pilot episode of *The Colbert Report*, a television show that aired on Comedy Central from 2005–2014 satirising personality-driven political pundit shows on Fox News, comedian Stephen Colbert introduced the term *truthiness*. Rather than leaning into the uncomfortable realities of truth (not even necessarily capital-T Truth), truthiness instead relies on the *feeling* (or the emotional appeal) of something being fact, even if it's fiction. Truthiness has functioned in late-night American television as satire and parody and as a tool to critique politics and news media, from talk shows like Colbert's to variety comedy shows like *Saturday Night Live*. While Colbert coined the term in a satirical way (quoted above), he saw the acceptance of truthiness (or narratives that *feel* real) in society as a precursor to authoritarianism (Rabin 2006). Indeed, though Colbert was satirising right-wing pundits in the US, truthiness was never satire; it was a serious indictment of American news media and politics. On *The Late Show with Stephen Colbert* (2016), he noted that "post-truth", a term coined to address how opinion based in emotional appeals was more influential than objective facts (circa Donald Trump's first presidential campaign), was just a "rip-off of truthiness".

Broken down to its basic principles, truthiness functions as a tool to build narratives and realities that *feel* real, even if they are not. Within video games, truthiness has been used as a narrative-building device to situate players in a version of history that is easy to digest and is not too uncomfortable. It's perhaps easy to see why video game franchises that rely heavily on historical events to inform narrative would employ truthiness rather than truth to shape narratives and build worlds. Truthiness is entertaining and, for some, safe. Military shooters, like *Call of Duty*, utilise truthiness (or emotional appeals grounded in something that *feels* real) to shape narratives that place the player in a heroic role. After all, who doesn't want to *feel* like the hero?[1] As a result, *Call of Duty* has played fast and loose with historical events. To cite an infamous example, in *Call of Duty: Modern Warfare* (Infinity Ward 2019), the Highway of Death massacre[2] is attributed not to the United States, but to Russia. Matthew Gault (2019) reflects on the ease of historical revisionism in games, noting, "[j]ust like that, in a piece of popular fiction, American atrocities become Russian atrocities. It's the kind of distance players need to separate themselves from history." (n. p.) In this retelling, the US (and by extension, the player) is the hero and history is revised for mass-consumption by an audience who may have no historical recollection or knowledge of the event. It's also here, situated in this retelling, that genocidal actions are a safe playground for players who are then able to (re)view themselves and the US as heroic based on the revisionist history that Matthew Payne (2012) asserts "lessens the potency of any critical protests against *Call of Duty*'s representation of contemporary war" (323). War becomes abstract, something that happens *over there*, away from the game space.

Our primary concern with this article is how historical events and their associated cultural periods, such as the Cold War, are re-visioned[3] not just in video games, but in video game paratexts[4]. In this article, we examine the first three

---

1   This is tongue-in-cheek. There is a notable link between war games, such as the *Call of Duty* franchise, and American exceptionalism. *Feeling* like a hero doesn't necessarily make one a hero, and spinning historical events is a part of American exceptionalism.

2   Holsti (2011): "The Iraqi invaders began retreating from Kuwait within two days, setting fire to oil fields as they left. Air attacks inflicted heavy casualties on retreating forces along what became known as 'the highway of death'. American, British, and French units pursued the Iraqis to within 150 miles of Baghdad. At that point, one hundred hours after the start of ground operations, President Bush ordered a cease-fire, and on April 6 he declared that Kuwait had been liberated." (20)

3   Here we use 're-visioning' to denote not only the sense of being revisionist, but also in the sense of reviewing history through the lens of truthiness/inaccuracy.

4   For the purposes of this article, we are defining paratexts as texts (regardless of medium) that are tangentially or directly connected to another text (in this case video

trailers[5] released for *Call of Duty: Black Ops 6* (Treyarch/Raven Software 2024): *wake up (rushmore)*, *Open Your Eyes*, and *The Truth Lies*. We have chosen to focus on these trailers because they precede the release of the game and are situated in such a way that they are meant to set the political and narrative stage for *Black Ops 6*. It's the trailer that John Ellis (1992) suggests "[...] is considerably more expensive per second than the programme it fronts; it's highly organised and synoptic, providing a kind of narrative image for its programme" (120) (an argument that was extended to movie trailers by Gérard Genette in 1997). Stephen Heath (1977) described the context of films, noting that our experience and understanding of a film extends beyond watching, encompassing our experiences before, during, and after, including historical knowledge, cultural knowledge, collective knowledge, personal experiences, and marketing. While Ellis, Genette, and Heath's early theories are speaking specifically of television and movie trailers, current media scholars have built upon this to both analyse video game trailers as cultural texts and to increase understanding of how game trailers have moved past their original purposes of showcasing hardware and gameplay (Švelch 2015)[6]. Beyond the established universe of *Call of Duty*, the trailers for *Black Ops 6* are the player's (and viewer's) first introduction to the game's world. While games that utilise historical events in their narratives offer the opportunity to engage with and build interest in history, especially in populations that have no prior interest (and little prior knowledge) of the historical period, the blending of actual historical references and 'deep fakes'[7] ultimately reinforces the truthiness of this revised history. For some, truthiness may satisfy the desire for a truth that does indeed reinforce their beliefs, like being of a superior race; in some ways, truthiness eases the need to search for truth. We have structured this article in sections that follow each trailer's release; like the trailers, the sections build on one another to form a complete picture through close readings and analysis, political and historical implications, and discussion. We analyse these trailers through the lenses of para-

    games) but are created independently of the video game. We are following Seiwald and Vollans (2023) consideration of paratexts as "textual 'other'" (4).

5   The scope of this project is limited to these three trailers in particular because there was simply not enough room to discuss the social media campaigns, PR materials that were sent to content creators, livestream analyses (including our own), and the *Black Ops 6* website (including the earlier versions that are no longer accessible). The original version of these trailers are, at the time of this writing, still accessible.

6   Cf. Ellis (1992), Gray (2010), and Villegas et. al. (2023).

7   A deep fake/deepfake can be defined as a realistic cinematic artefact of an individual who has been digitally restructured using advanced visual effects, often with the intention of spreading disingenuous information. With the emergence of AI, deepfakes are getting increasingly difficult to identify. In addition, it's an emergent field of scholarship (see Westerlund 2019; Hazan 2020; Murphy/Flynn 2022; Tariq/Abuadbba/Moore 2023; Doğan Akkaya 2024).

textual theory, truthiness, and Cold War historiography in order to explore and examine the problems and potentialities of using actual historical accounts in the processes of worldbuilding in game IPs like *Call of Duty*. We ultimately argue for increased responsibility in both industry and academia to address how truth and truthiness are used in video games to build narratives that revise historical events and cultural periods into stories that *feel* real – especially when the lines between reality and virtual are blurred.

This article takes a non-standard form that integrates lyrics that both draws upon songs that have been representative of previous games in *Call of Duty* and builds upon *Black Op 6*'s narrative and version of Cold War history that it creates. It has been the norm for *Call of Duty*'s official trailers to feature the music of well-known musicians, a creative choice meant to set the tone for the game and build anticipation for the game's release among the audience of gamers for which the music is intended. Past releases of games set in the *Call of Duty* universe have featured the music of musicians such as Metallica (*Call of Duty: Modern Warfare*, Infinity Ward 2007), Eminem (*Call of Duty: Modern Warfare 2*, Infinity Ward 2009), The Rolling Stones (*Call of Duty: Black Ops*, Treyarch 2010; *Call of Duty: Black Ops III*, Treyarch 2015), AC/DC, (*Call of Duty: Black Ops II*, Treyarch 2012), Frank Sinatra (*Call of Duty: Ghosts*, Infinity Ward 2013), Guns N' Roses (*Call of Duty: Infinite Warfare*, Infinity Ward 2016), New Order (*Call of Duty: Black Ops Cold War*, Treyarch/Raven Software 2020), The Notorious B.I.G (*Call of Duty: Black Ops Cold War* 2020), and 21 Savage (*Call of Duty: Modern Warfare III*, Sledgehammer Games 2023). *Call of Duty: Black Ops 6* (2024) was no different, with the official trailer featuring *Detonator* by Killer Mike, a politically outspoken rapper whose views have, according to some political pundits, had some significant sway in recent US elections (McLaughlin 2020; Ramsey 2020). Music has been central to the trailers and marketing for *Call of Duty* games; it's for this reason that we have also chosen to frame each section of our article with specific song lyrics.

2. Background and context

> And now you want to trump me
> Prison is a business, America's the company
> Investing in injustice, fear and long suffering
> We staring in the face of hate again
> The same hate they say will make America great again
>
> "Letter to the Free," Common

As researchers in the United States, it's important that we address the backdrop of this article: that of truthiness, the current American political landscape, and video games like *Call of Duty*. The heart of this article is wrapped in writing sessions and discussions spread across a short, turbulent timeline that is influenced, in large part, by truthiness. By circumstance, we are forced to reckon with the collision of our goals in 2024 and our realities in 2025. To put it plainly, there is something markedly different (and difficult) about writing on *Call of Duty: Black Ops 6* (2024) in 2024 versus 2025. The trailers we discuss in this article were released in May 2024. Following a livestreamed analysis of these trailers on Twitch (Blackmon 2024), we proposed this article in August 2024. Coinciding with the October 25, 2024 release of *Call of Duty: Black Ops 6*, we released a podcast on *Call of Duty* and social critique against the backdrop of a chaotic election cycle, the feeling of hope crowded by the reality of American sentiment towards race and gender (Blackmon/braegger/Lukomski 2024, October 25). In the liminal space between election day and inauguration day, we released a podcast on the *Call of Duty: Black Ops 6* campaign (Blackmon/braegger/Lukomski 2024, December 18).

We completed this article in early-2025. Donald J. Trump's second term as president of the United States is shaping not the rise (as that has already happened), but the increasing hold of fascism over the US. Recent headlines include immigrants (and citizens) deported without due process (Walters 2025; Bustillo 2025); slashes to government funded research (Bendix 2025); government corruption (McNicholas/Poydock 2025; Dickinson 2025); and the destruction of the social safety net (Cassidy 2025). The list grows longer each day. While responding to feedback and completing revisions for this article, two significant events happened: Trump launched an attack on Iran without Congressional approval, using justifications (propaganda) that echoes that which was used twenty years ago during the War on Terror, and Congress (both the House and Senate) passed a bill that representatives and senators acknowledged was horrendous for the middle-class and poor while still voting for it. During final revisions, CBS ended *The Late Show with Stephen Colbert*, effective May 2026. This came just days after Colbert criticised a settlement between Trump and CBS's parent-company, Paramount Global (Bauder/Rancilio/Dalton 2025). This is truthiness in action.

As researchers who focus on diversity, equity, inclusion, and accessibility (also referred to as DEI, DEIA, or IDEA) in video games, our research has been targeted in the name of *truthiness*. Universities are losing funding if they don't bend to the whims of the current administration. Pushback through the judicial system is slow; the damage will be long-lasting. To put it plainly, discussing *truthiness* and *truth* has a different impact *now* when our reality in the US is shaped by *truthiness*, spread by a few and embraced by millions. Truthiness has serious repercussions. In analysing the narrative of *Far Cry 5* (Ubisoft 2018), Piero (2024) calls for increased industry responsibility when deploying right-wing narratives within games, as well as greater critical engagement and 'bold refutation' amongst game studies scholars in response to these narratives. We contend that paratexts – such as those released for *Call of Duty: Black Ops 6* – that re-vision history receive the same critical engagement and analysis. When history is re-visioned for those with little experience or knowledge of the topic, the images presented become accepted truth.

3. Wake up: Setting the stage

> Everybody's going to the party
> Have a real good time
> Dancin' in the desert
> Blowin' up the sunshine
> Blast off, it's party time
> And we don't live in a fascist nation
> Blast off, it's party time
> And where the fuck are you?
>
> "B.Y.O.B.," SYSTEM OF A DOWN

In his analysis of marketing paratexts for post-9/11 military shooters, particularly *Call of Duty 4: Modern Warfare* (2007), Payne (2012) describes these trailers as "overwhelmingly concerned with selling only select elements of military realisticness: sophisticated enemy artificial intelligence, military weapons and vehicles that function and look like the real thing, and combat that unfolds in authentic theatres of war" (310). This focus on gameplay footage as an element of military realisticness highlights innovation in the game engine, as well as equipment the players will use and the stunning visuals of the game environment is apparent in trailers for nearly every successive *Call of Duty* game. In the 2019 remake of *Modern Warfare*, the official reveal trailer started with the text, "[a]ctual in-game footage" (Call of Duty 2019), before a voiceover narrates, "the rules have changed. There's a fine line between right and wrong" (ibid. 2019). What follows the voiceover is gameplay footage and first-person perspectives that showcase the guns used by the player and visuals of the game's battlefields, including cityscapes and nondescript countries in the Middle East, a favorite of *Call of Duty*. Indeed, several *Call of Duty* games invite players to a safe and quasi-historical 'party' in the desert.

However, in recent years, there has been a shift in the trailers for Treyarch's *Black Ops* storyline. The initial trailers for *Black Ops* follow the established pattern. The reveal trailer for *Call of Duty: Black Ops II* (2012), set to the musical backdrop of *Ultraviolence* by Cliff Lin, features an aging Frank Woods, one of the primary characters for the series, as he criticises the focus on technological warfare in 2025; Woods ends the trailer by telling the viewer, "they'll always need men like us: those who are willing to do what others cannot" (Call of Duty 2012). Much like the trailer for *Call of Duty 4: Modern Warfare* (2007) analysed by Payne (2012), this trailer focuses on gameplay footage, cutscenes, and popular characters who push the narrative forward. As Dunne (2016) describes, video game trailers "[extend] the narrative beyond the players' experience of the games" (274). In the trailer for *Black Ops II* (Call of Duty 2012), Frank Woods sets the narrative tone for the game, moving the game into future technology.

As the release date for *Call of Duty: Black Ops II* (2012) neared, IGN (2012) released[8] the *Surprise* trailer, a live-action celebrity- and influencer-filled advertisement set to AC/DC's *Back in Black* that begins in a war-torn Los Angeles. Danilo Di Julio aims a sniper rifle at Zach Steel before iJustine throws a tomahawk near his head and exclaims, "[s]urprise!". The trailer carries on in this way with celebrities and influencers disrupting each other's shots, capitalising not only on the celebratory nature of a surprise party, but also on the height of the Marvel's Cinematic Universe by using Robert Downey Jr., who appears in a jet before saying, "[g]uess who brought a jet to a gunfight?" Zach Steel, having survived due to the disrupted sniper shot, then triggers a hellfire missile using a portable radar strapper to his wrist, ending the battle. As he walks away in slow motion from the explosion behind him, the text overlay reads, "[t]here's a soldier in all of us." While this trailer didn't showcase in-game footage, it did showcase weapons, equipment, and scorestreaks the player would be able to use in the game (sniper rifle, tomahawk, autonomous ground robot, lightning strike, and a hellfire missile, among others), as well as a representation of players *playing* the game in a multiplayer environment. As such, it's evident the trailers for *Call of Duty: Black Ops II* (2012) adhered to the familiar formula seen in the trailers for previous titles. These trailers showcase footage and mechanics and, we would argue, are *purchase-coded* trailers, or trailers that focus primarily on selling games by foregrounding mechanics and new (or returning) features over story and narrative.

Purchase-coded trailers for video games are expected. At their core, video game trailers are marketing tools designed to generate excitement in consumers. The hype cycle in the games industry tends to play out predictably: a game is announced a year or two before its release with a short trailer. Following the release, journalists report on the trailer while fans dissect each frame, with the comment section for the trailer becoming a hub for engagement. This spreads onto other internet forums, such as Reddit. Each new trailer offers new pieces to add to the puzzle, the excitement (or hype) growing with each new addition. For example, *Red Dead Redemption 2* (Rockstar Games 2018) was announced on Rockstar Games' (2016) Twitter account on October 18, 2016, the accompanying image noting a trailer would be released on October 20, 2016. IGN (2016) then released the one-minute trailer on October 20, 2016, sparking discussion in the comments and prompting posts on Twitter and Reddit. Though the game was expected in Fall 2017, it didn't release until October 26, 2018. In the two years leading up to the release of *Red Dead Redemption 2*, Rockstar Games unveiled trailers, interviews, and websites dedicated to the marketing and development of

---

8 IGN co-releases official Call of Duty trailers.

the game's world. These paratexts contributed to and drove the hype cycle for the game for two years; Wright (2018) notes that these paratexts invited viewers to engage in a shared reconstruction of history. However, digital paratexts for video games (and thus the hype cycle surrounding the games) are often difficult for researchers to engage with post-release, as components are lost to linkrot, link re-use and re-direction and the outright removal of materials. This is described by Wright in their discussion of *Red Dead Redemption 2*, as blog posts about the use of history within the game's development disappeared from Rockstar's website post-release. It's something we encountered while writing this article, as well. We had intended to include the website co-released with *The Truth Lies* advertisement campaign, which had featured an old television that could, upon clicking, switch channels. These channels unlocked with the trailers and provided additional paratextual engagement for the audience. However, in October 2024, the website changed to *The Truth Dies* when Activision switched to marketing *Black Ops 6's* zombies campaign. The link to the website was re-used and the online internet archive did not periodically archive the page's contents. Digital paratexts present unique challenges; while finding trailers is simple enough, paratexts such as websites, development blogs, and even VODs[9] from streamers who engaged with the paratexts during release often succumb to linkrot.

However, the *Call of Duty* hype cycle is mitigated by their annual release schedule, which minimises the need for purchase-coded trailers. Since 2003, Activision has released a new *Call of Duty* game each year, with primary development splitting between Infinity Ward, Treyarch, and Sledgehammer Games. In other words, players *expect* a new release each year and, as a result, the games frequently appear on best-selling and top-grossing charts (note that we are not saying these games appear on the best-*rated* lists). This allows wiggle room for paratexts promoting the game. Vollans (2023) writes, "[t]railers are not wholly or just promotion, but articulations of a product, of 'what could be', even if this articulation is inaccurate" (161). Foregoing the need to explicitly rely on purchase-coded trailers in recent years has an additional side effect: the trailers are not only playful, but set the cultural, historical, and political context for contemporary entries in the franchise. Activision's version of the Cold War is written not only with the re-visioning of history (truthiness in action), but also with the strategically released trailers and paratexts

The first trailer for *Call of Duty: Black Ops 6* was released on May 23, 2024. [W]ake up (rushmore) #thetruthlies is a 42-second found-footage style video. Everything happens from the first-person perspective of the viewer, placing them directly in the action. Gravel crunches underfoot as a spotlight shines on the four

---

9   Videos on demand, or VODs, are stream recordings from platforms such as Twitch and YouTube. Unless downloaded and re-uploaded, Twitch removes VODs after a set amount of time.

presidential faces of Mount Rushmore; a disembodied voice intones, "...should wake 'em up". The small group in the video make their way through the dark, climbing their way up the rockface and laying down black tarp, every video clip unsteady and rushed, each actor breathing heavily. After a screen blackout and cut, helicopters are heard in the background while the camera shakily focuses on Mount Rushmore, each of the faces' eyes now covered by a black tarp with orange writing. The tarps on George Washington, Thomas Jefferson, and Theodore Roosevelt read, "[t]he truth lies", while Abraham Lincoln's tarp is covered with a wolf symbol. Were it not for the video's release on the *Call of Duty* YouTube channel, the viewer would likely not know what the video was about; the sole context clue is the distinctive orange on black color scheme customarily used in the *Black Ops* series.

Though the trailer is short, it provides the first historical and cultural grounding for the game. The found footage style of filming, though initially used for horror films in the 1960s and 1980s, was revived and brought to mainstream media focus by *The Blair Witch Project* (1999). By incorporating a found-footage aesthetic, this trailer connected *Call of Duty: Black Ops 6* to both the horror genre and to 1990s American culture, establishing a nostalgic approach to the game that appeals to both personal nostalgia (personal memories and experiences within a period) and historical nostalgia (impressions – not personally experienced – of how the period was) (Marchegiani/Phau 2010; Stern 1992). While nostalgia is an effective tool for video games that incorporate historical themes, it should be noted that this appeal to nostalgia may be for a past that does not necessarily exist (Bowman/Wulf 2023; Muehling/Pascal 2011; Fenty 2008; Makai 2018). This distinction is key to re-visioning history; nostalgia, an emotional response, plays well with truthiness.

Though more evident in later trailers for the game, this trailer makes appeals to those with distinct memories of the 1990s while establishing a re-visioned historical setting for the game. The trailer taps into 1990s American culture in a way that appeals to those who have memories of the 1990s, while at the same time, may be wary of the modern political system. Wright (2023: 33) argues that games that tap into beloved cultural facets, such as *The Blair Witch Project*, create an intertextual link that ties the game to cultural touchstones that encourages players to engage with the past in predescribed ways. At the same time, trailers are snapshots of the environment in which they are produced (Vollans 2023; Wright 2023). That is to say, the first trailer for *Call of Duty: Black Ops 6* (2024) sets the stage for a nostalgic approach to 1990s American culture that utilises contemporary politics by insisting to the viewer that 'the truth lies', a political statement that, if not obvious enough, is posted over the four presidential heads of Mount Rushmore. The trailer establishes a re-visioned historical setting for both players who are nostalgic for the game's historical period, as well as those who have little to no memory of it.

4. Open your eyes: Blending fiction with fact

> Proud to be a Black man
> Livin' in the land of the brave and the free
> Yeah, I'm all-American
> And that American dream ain't cheap
> We've come a long way
> Still got a long way to go
> When you're livin' as a Black man
> It's a different kinda 'merican dream
>
> "AMERICAN DREAM," WILLIE JONES

The second release trailer for *Call of Duty: Black Ops 6* (2024), *Open Your Eyes*[10] (IGN 2024), blurs the lines between a specifically crafted reality and fictional gameworlds by blending real archival footage with game narrative. As such, the version of the Cold War presented in the trailer has just enough truthiness to it to simultaneously feel realistic and be safe from any actual historical interrogation. Payne (2012) makes the assertion that,

"The advertised pleasures of playing wars past, present, or future is, in actuality, the pleasure of playing with a delimited textual realisticness, and not a contextual realism that connects the gamer and game to the lived realities of an outside world. Video game marketing of commercial military shooters largely works to collapse the divide between textual realisticness with any broader understandings of "realism" to argue that their game's attention to technical detail offers the necessary representational and simulational bona fides to engender an immersive reality available to any who might buy their electronic wares." (Payne 2012: 309-310)

In expanding the assertion to paratexts, Seiwald and Vollans (2023) describe how historical games offer players "the chance to revisit past events without having to fulfil the impossible task of traveling back in time to those events" (9) and, as a result, the associated paratexts (from trailers, to websites, to artwork, to purposeful links) "shape popular understandings of specific periods, events, and people" (ibid.). By blending reality with fiction, Activision presents an image of post-Cold War America within the paratexts of *Black Ops 6* that is safe to play in.

Open Your Eyes is a montage of quickly cycling clips, overwhelming the viewer with historical and cultural information about the 1990s. The trailer opens with

---

10 'Open your eyes' has ties to alt-right communities in the United States, as well as to online alt-right communities (red pill). While the connection is certainly one to be noted, it's outside the scope of this article and could be a place for future study.

the familiar scratching noise of a VCR rolling a VHS tape, the static giving way to a Chia Pet advertisement from the early 1990s, the happy staccato "ch-ch-ch-Chia!" echoing behind the static overlay. The clip cycles. The Chia Pets are replaced by a special bulletin reading, "1991 Financial Report". The familiar, "[h]ave fun and watch it grow!" turns into a newscaster reporting, "[m]edian home prices have fallen to $120,000". The clip starts with an aerial shot of a suburban neighbourhood; before the announcer can finish, the clip cycles to a close-up of an intersection, the streets labelled Pleasant Valley and Tranquility Lane. The video cycles to Bill Clinton playing the saxophone before quickly cycling to another special bulletin, the announcer telling us, "[r]ecruitment for the CIA is at an all-time high" as video clips of the Pentagon and the Washington Monument play in the background. The video cycles again to a happy couple walking on the beach, the TV show's name in notebook-style font reading *Frank's Woods*. The announcer's cadence is familiar to those who watched teen dramas in the late 90s and early 00s, telling us, "[n]ew episode, tonight at nine!". We are fifteen seconds into the second trailer for *Black Ops 6*, each cycle punctuated by the loud static familiar to those who changed channels on CRT TVs, and more than half of the clips are fabrications.

The first clip – an advertisement for Chia Pets – is real. The advertisement is available on YouTube, posted by the appropriately-titled channel *Retropond* (2021). In the *Black Ops 6* trailer, the image is washed out and mirrored in spots (adding to the vintage aesthetic); the Chia Dog, once facing right, now faces left. The third clip, Bill Clinton playing the saxophone, is also real; there is no shortage of footage of Clinton playing the saxophone. That leaves us with the remaining three clips of the first fifteen seconds: the 1991 Financial Report, the CIA Recruiting Report, and the ad for *Frank's Woods*. The final clip – *Frank's Woods* – is easy to spot as fake; it's an easter egg for fans of the *Black Ops* storyline, named for one of the main characters, Frank Woods. The final two clips – the special bulletin news reports – *could* be real. They're positioned in such a way that their validity is hard to confirm; brief special bulletins from news cycles in the early 1990s are hard to track down without knowledge of the local news station or date. It's knowledge of the game – knowledge that comes after the game's release – that confirms these clips are fake. Tranquility Lane, noted in the first clip on median home prices, is a reference to the zombies campaign in *Black Ops 6*. At the time of the trailer's release, nothing was known about the game's zombies campaign narratively. The CIA recruitment clip is a reference to the main campaign in *Black Ops 6*, which, at the time of the trailer's release, was also difficult to discern.

After these clips, an emergency alert tone sounds, followed by a burst of static that signals a switch in narrative. The clips become shorter, cycling nearly every second as the trailer bombards the viewer with information. A narrator directly addresses the viewer, greeting them with a voice changer commonly heard in videos produced by online hacktivist group, *Anonymous*. The narrator's message is composed of their own modulated narration and snippets of political figures speaking. For example, the opening to this part of the trailer features Margaret Thatcher, George H. W. Bush, Al Gore, Bill Clinton, and an unidentified person[11], their clips blended with the narrator occasionally echoing their words (indicated in parentheses):

Narrator: Hello (hello).
Thatcher: Ladies and gentlemen,
Bush: we are talking
Gore: to
Unidentified: you
Narrator: We are talking to you (to you) You.
Bush: You may remember (remember).
Clinton: a better time (better time).

The clips in this section of the trailer are primarily real and easy to find: political figures giving speeches and interviews, burning oil derricks, traffic in LA, night-vision missile launches, and dilating eyes. The first in-game footage appears at 0:35, of a helicopter taking off while an oil derrick burns behind it, and is followed by a blink-and-you-miss-it timer reading "0:0B:06:#" that is counting down, a reference to the game that is reminiscent of the numbered codes associated with the *Black Ops* series. The combination of the footage and timer lasts until 0:37. Uniting this section of the trailer are clips from what we can only describe as a patriotic 1991 internal training and promotional video for Sizzler, a Californian surf-and-turf fast-casual steakhouse associated with *fancier* middle-class dining (Redhouse 2012). Clips from the training video that were used in the trailer include couples walking down a sunny boardwalk, men looking at blueprints for a building and nodding confidently at the camera, children playing baseball, people in business clothes, filling their plates with salad from an all-you-can-eat salad bar, and groups smiling and laughing over food.

The clips from the Sizzler training video provide a glimpse into 1990s American culture that portrays a stylised version of the American dream: happy families with expendable income that can be used for extracurriculars and food. This is a time of comfort and excess. What sets the Sizzler training video into unabashed patriotism is the song used throughout the video. When the training video starts, the music swells and a chorus sings the first verse:

---

11 Rather than an easily identifiable political figure, the unidentified person's identity is difficult to discern. He appears for half a second at 0:22 in the trailer for the word 'you'. He doesn't appear again. Clips for this trailer were pulled from the late 1980s to the late 1990s, and the brief glimpse provides very little information about his identity.

All across America, the song of freedom rings / The song is growing stronger every day / It tells us when we listen to the message that it sings / Let us lift our voices / We can make the choices / We will make the most of all the best that freedom brings / Sizzler is the one who brings us choices / Reaching out across the USA / Each and every day, get a little freedom in your life.

While the *Black Ops 6* trailer doesn't lift the entire song, it does lift two refrains: "Let freedom ring!" and "...the USA!" The first refrain, "Let freedom ring", is most commonly associated with, *My Country, 'Tis of Thee*[12], a patriotic song that served as the American national anthem before *The Star-Spangled Banner* was adopted officially. The brief clip from the Sizzler ad, slightly distorted, plays at the end of the first clipped together section of the narrator's message noted above, punctuated with a child hitting a baseball as the refrain plays. After the brief musical refrain, the narrator comes back, stating, "[b]ut it was false", before George H. W. Bush clips together, "the truth lies". The second refrain, "...the USA", occurs at the end of the trailer. It immediately follows a clip of Bill Clinton stating, "[b]lessed are the peacemakers", a quote from the Bible he used in his 1995 speech in Northern Ireland supporting peace progress, and is played over a clip of a missile hitting a target in night-vision.

The second part of the trailer uses the Sizzler video to paint an image of 1990s American culture, interspersed by world leaders who disrupt the harmony of the video. Where the first section of the trailer plays *Two Truths and a Lie* with the viewer, the second section breaks down the viewer's knowledge of the historical period, telling the viewer their historical memory and historical knowledge is a lie by using the voices and imagery associated with the period to make the case. Sizzler, a staple of 1990s American culture, ties together the American image of family and economic prosperity, bolstering the clips of positive economic news in the first section of the video. Sizzler is *fancy* in a way that is entrenched in American culture and is something to be attained: an abundance of food, bright interior lights, warm colors, and happy families. Cecilia Hae-Jin Lee (2015, n. p.), an immigrant who moved to the United States as a child with her family, described the restaurant as "the epitome of the American meal". However, by the end of the trailer, the viewer is left with the understanding that this display of American history and culture is a lie. Simply put, this truth – as the viewer understands or remembers it – is a lie.

The second trailer for *Black Ops 6* creates a picture of an American Dream that never was, building upon memories of media and feelings of nostalgia that

---

12 *My Country, 'Tis of Thee* utilised the British national anthem's melody and new lyrics. The first verse is where *Let freedom ring* appears: My country, 'tis of thee / Sweet land of liberty / Of thee I sing / Land where my fathers died / Land of the pilgrim's pride / From every mountainside / Let freedom ring!

are meant to reinforce the understanding that while the 'American Dream' may not be cheap politically, it's definitely worth going to war to protect. After all, as Clinton intones, "[b]lessed are the peacemakers" while a patriotic hymn sings, "... the USA", all as a missile decimates a target. Each clip builds a re-vision of history that is set both in reality and in the gameworld. Indeed, throughout the short trailer, there are more 'real' clips than those from the gameworld itself. By interspersing historically and thematically related clips that were real and fictional, the first two trailers not only set the tone for the game's narrative, but also created a re-visioned tone for the period of the 1990s. This, in essence, establishes the truthiness of the second trailer and the narrative that is being constructed, both by the trailer and the game itself. This truthiness is deepened with the release of the third trailer that relies not only on the use of nostalgia, but also on CGI and Deep Fakes to further obfuscate the history of the immediate post-Cold War period (or Activision's version of it) that the game is using as the foundation of its narrative.

5. The truth lies: CGI, deep fakes, and the uncanny valley

> Evil leader, let the evil by other evil leaders
> We pray for peace and they give you war, pray for more
> You begging, then back some more, and they just ignore
> They bag you up for the score, and then they shut the door
> No empathy or pity for you anymore
> Already in Hell, formed in fire from those wars
> The empires, we are led by them liars
> Wake the fuck up right now, 'cause times is gettin' dire
> Detonator
>
> [Featuring Rock D., the Legend], Killer Mike

The third trailer premiered on May 28, 2024, five days after the second trailer, and is entitled *Black Ops 6: 'The Truth Lies' - Live Action Reveal Trailer*. However, this trailer is anything but *live action*. Rather than 'live action' clips (or clips from the gameworld), what we get are hyper-realistic, deep-fake quality computer generated depictions of then-US presidential candidate and Governor of Arkansas Bill Clinton, US Army General and Chairman of the Joint Chiefs of Staff[13] Colin

---

13   It's important to note here that at the time that these events were taking place historically, Colin Powell was the chairman of the Joint Chiefs of Staff (CJCS) because he would have not only have been the highest ranking military officer in the United States, but he would have also have been the primary military advisor to the president of the United States. His words, his advice, held as much (if not more) weight

Powell, former UK Prime Minister Margaret Thatcher, US President George H. W. Bush, and President and Prime Minister (dictator) of Iraq Saddam Hussein in a montage sequence where they all seem to be finishing each other's sentences and ultimately telling the story of the global political situation surrounding the Gulf War.

Clinton opens the trailer, telling the viewer, "[t]he truth is, your whole life is a lie"; Powell discloses, "[y]ou wanted peace, so we hid the war. Now it rages all in the shadows" (ostensibly granting the 'peace' that the people of Killer Mike's America have been praying for); Thatcher adds, "behind closed doors where we hide the world's secrets"; H. W. Bush (then-sitting US president) continues, "while we build your trust the world falls apart around you, if you weren't so distracted you'd realize..."; and Hussein concludes "nothing is what it seems. But if it's truth you seek, look in the dark". Uninterrupted, the full speech reads:

The truth is, your whole life is a lie. You wanted peace, so we hid the war. Now it rages all in the shadows behind closed doors where we hide the world's secrets. While we build your trust the world falls apart around you. If you weren't so distracted you'd realize nothing is what it seems. But if it's truth you seek, look in the dark.

It's only when we combine all of the snippets from these 'world leaders' – allies and enemies alike – that we get the full message, the full story of the political situation. What is most interesting about this message is that it does not, in fact, come from a live action trailer, nor in pieced-together clips of the leaders themselves. Rather, the message comes from deep fake quality depictions of these people – a fact that serves to blur the line between fact and fiction and gives credence to the truthiness of the message. The appeal is emotional, designed to raise a response in the viewer.

This line between fact and fiction is blurred even further by the real footage of burning oil fields, night vision shots of helicopters, and flashes of the coordinates of Saddam Hussein's presidential palace that are interspersed with the CGI depictions. The juxtaposition of the 'real' and the deep-fake makes it nigh impossible to discern which is which, where one story ends and another begins, even for those who have studied and lived through these moments. Knowing history (and understanding which clips could be real) still does not help *not quite knowing* and questioning where the information is coming from. Video clips are ephemeral and source-checking is near impossible; this is further complicated when historical documents are made inaccessible[14]. The history – the mundanity of creating a

---

than anyone else's when it came to military movement and strategies in the United States.

14 Within the United States, government websites, physical archives, and museums are being scrubbed of art, artifacts, exhibits, data, and information that complicates

world and a culture mirroring a reality – is difficult to parse through. Beil (2021) notes:

"(Ludonarrative dissonance) characterises most video games that make use of historical settings. History becomes an assemblage of scenarios and props that primarily serve to embellish the story world. Thus, these games do not offer "performatory challenges", historical knowledge is not required for the successful completion of the game. History "takes place around and above the players, but their experience of history is fragmented, ontological and particularised." (58)

In this situation history and historical knowledge are not only not necessary for the consumption and understanding of the game's narrative, but in its practice of truthiness the game creates and imparts its own quasi-historical narrative that for some can be mistaken for history. In this moment there is no difference between CGI Colin Powell telling us that the war was hidden because Americans wanted the illusion of peace and the actuality of the war being portrayed as a series of night vision missile strikes on the evening news in a way that (felt more like) watching the 1983 film *War Games* than watching real people and real places being destroyed before our eyes.

6. Conclusion

> Here's to the greater good, for all
> Do what you know you should, for all
> We all may die
> Something's going on, just look around
> Fear is on the rise and there's blood all over the ground
> Let's all just blindfold the poor, we must remind them what's in store
> We got 'em now, just break 'em down a little bit more
>
> "Feed the Machine," Poor Man's Poison

Notably, the historical context for *Call of Duty: Black Ops 6* – set during the First Gulf War's Operation Desert Storm in 1991, amidst the collapse of the USSR – is minimised in the game itself. What the United States Office of the Historian (n.d.) describes as the 'first full-scale post-Cold War international crisis' does not get built out in the game's narrative. Rather, the game takes place in a familiar *Call of Duty* environment (the Middle East), with familiar *Call of Duty* missions (relatively

---

particular versions (white heterosexual male) of history (Yourish/ Daniel/ Datar/ White/ Gamio 2025; Reuters 2025; Singer 2025; Mauran 2025)

linear run-and-gun maps while the player runs from checkpoint to checkpoint). Instead, the historical context for *Black Ops 6* is provided primarily via paratexts that rely on truthiness to build a re-visioned history. Historical figures from the trailers such as Colin Powell, George H. W. Bush, and Margaret Thatcher are not featured in the game (though their actions and the consequences of those actions certainly are). Saddam Hussein is mentioned in passing and in environmental settings, as his palace is a mission location. Bill Clinton does appear incidentally in the game, at a Clinton campaign rally that serves as a mission location; the player's goal is to take a picture of a fictional senator who is standing next to Clinton.

The trailers for *Call of Duty: Black Ops 6* build Activision's re-vision of the Cold War (and immediate post-Cold War) through cultural intertextuality that has little resemblance with the game's narrative itself. Montages of 1990s American culture (both real and imagined) are clipped together with political speeches (both real and imagined). At each stage, the viewer is told not to trust the government and to accept the truthiness of the game. *The truth lies*, also stylised as #thetruthlies for social media marketing purposes, is the fundamental core of the game, wherein the main characters are constantly confronted with a falsehood they had believed to be truth. The predominant theme is that the government and the politicians are lying to the watcher of the trailers and, ultimately, the player who plays the game. For example, the main character, Troy Marshall, contends with corruption within the CIA and his loyalty to the organisation and the government. The image of the American Dream built through the paratexts is even further problematised and complicated by the positioning of Marshall, a Black man, as the protagonist. Not only is it problematic that he is positioned (initially) as being unquestionably loyal to America and its cause, but he is ultimately turned on and turned out by the country that he loves and that can and will never love him (openly) because it's ultimately corrupt. He can only serve his country in secret and can only save his country by fighting against its current power structures.

Truthiness – not truth – packaged in a re-visioned immediate post-Cold War setting is the underlying narrative theme of *Call of Duty: Black Ops 6*. However, using realism to create a narrative without critically engaging with the underlying theme – that the government is lying, as told by powerful figures in hyper-real CGI that could pass as deepfakes – requires further attention. There has been research on how media is used to influence people, especially young men, towards distrust in government and anti-intellectualism (Scharrer 2004; Shapiro 2010; Claussen 2011; Friedman 2017; Dagnes 2019; Lee & Hosam 2020); fear is, indeed, on the rise. And it's, ultimately, this fear that feeds the machine of American nationalism that underlies the truthiness we see in video games and their associated paratexts. Truthiness in the United States has had a very real impact, with the 2016 election of Trump based on claims of government conspiracy and lies (i.e. calls to 'drain the swamp') and claimed media complicity in lies (i.e. 'fake news' levelled at legacy news organisations). The truth, Trump claimed, was a lie; he provided a re-visioned history based on truthiness. While acknowledging the impact of truthiness on the American historical, political, and social landscape, we are calling for critical engagement and research that extends beyond video games themselves to the paratexts that utilise re-visioned historical narratives to both market games and create narratives.

References

Bauder, D./Rancilio, A./Dalton, A. (18 July 2025): "Stephen Colbert's 'Late Show' is Canceled by CBS and Will End in May 2026." Retrieved from https://apnews.com/article/stephen-colbert-late-show-cbs-end-8bad9f16f076df52coff-c50e9c8adbab.

Beil, B. (2021): "'And You Didn't even look at it!' Assassin's Creed (Self-)Discovery Tour." In: Beil, B./Freyermuth, G. S./Schmidt, H. C. (eds.), Paratextualizing Games Investigations on the Paraphernalia and Peripheries of Play. Bielefeld: transcript, pp. 58–74.

Bendix, A. (8 May 2025): "Trump Administration Cut More Than $1.8 Billion in NIH grants." Retrieved from https://www.nbcnews.com/health/health-news/trump-administration-cut-18-billion-nih-grants-rcna205568.

Blackmon, S. [Saffista]. (2 August 2024): "[Call of Duty and the Cold War] Black Ops 6 Trailers Deep Dive (No Gameplay) With @toskakoshka" [Video file].

Blackmon, S./braegger, v./Lukomski, J. (25 October 2024): "Episode 252: The Game is a Lie: Call of Duty and Social Critique" [Podcast]. https://www.nymgamer.com/?p=18586.

Blackmon, S./braegger, v./Lukomski, J. (18 December 2024): "Episode 254: The Truth is a Lie: Black Ops 6 Spoilercast" [Podcast]. https://www.nymgamer.com/?p=18597.

Bowman, N. D./Wulf, T. (2023): "Nostalgia in Video Games." Current Opinion in Psychology 49. Retrieved from https://doi.org/10.1016/j.copsyc.2022.101544.

Bustillo, X. (29 April 2025): "Trump Wants to Bypass Immigration Courts. Experts Warn it's a 'Slippery Slope'". Retrieved from https://www.npr.org/2025/04/29/g-s1-63187/trump-courts-immigration-judges-due-process.

Call of Duty (1 May 2012): "Official Reveal Trailer | Call of Duty: Black Ops 2" [Video file]. Retrieved from https://www.youtube.com/watch?v=x3tedIWs1XY.

Call of Duty (23 May 2024): "wake up (rushmore) #thetruthlies" [Video file]. Retrieved from https://www.youtube.com/watch?v=RxIHyfwUrTg.

Call of Duty (28 May 2024): "Black Ops 6: 'The Truth Lies' - Live Action Reveal Trailer" [Video]. YouTube. Retrieved from https://www.youtube.com/watch?v=Vo8UPqchVgQ.

Cassidy, J. (24 March 2025): "Don't Believe Trump's Promise About Protecting the Social Safety Net". Retrieved from https://www.newyorker.com/news/the-financial-page/dont-believe-trumps-promises-about-protecting-the-social-safety-net.

Claussen, D. S. (2011): "A Brief History of Anti-intellectualism in American Media." Academe 97(3), pp. 8–13. Retrieved from https://www.aaup.org/article/brief-history-anti-intellectualism-american-media.

Dagnes, A. (2019): Super Mad at Everything All the Time: Political Media and Our National Anger. Cham: Palgrave.

Dickinson, T. (3 June 2025): "Elon Musk's Reign of Corruption Chronicled in Elizabeth Warren Report". Retrieved from https://www.rollingstone.com/politics/politics-features/elon-musk-corruption-elizabeth-warren-report-1235353085/.

Doğan Akkaya, F. (2024): "Deepfake Dilemmas: Imagine Tomorrow's Surveillance Society through Three Scenarios." Journal of Economy, Culture and Society 70, pp. 121–134.

Dunne, D. J. (2016): "Paratext: The In-between of Structure and Play." In: C. Duret/C. Pons (eds.), Contemporary Research on Intertextuality in Video Games. New York: IGI Global, pp. 274–296.

Ellis, J. (1992): Visible Fictions: Cinema, Television, Video (Revised Ed.). London: Routledge.

ESA (2024): "Essential Facts About the US Video Game Industry". Retrieved from https://www.theesa.com/resources/essential-facts-about-the-us-video-game-industry/2024-data/.

Fenty, S. (2008): "Why Old School is 'cool': A Brief Analysis of Classic Video Game Nostalgia." In: L. N. Taylor/Z. Whalen (eds.), Playing the Past: History and Nostalgia in Video Games. Nashville: Vanderbilt University Press, pp. 19–31.

Friedman, U. (23 December 2017): "The Real-world Consequences of 'fake news'". Retrieved from https://www.theatlantic.com/international/archive/2017/12/trump-world-fake-news/548888/.

Gault, M. (29 October 2019): "'Modern warfare', The Highway of Death, and Call of Duty's Exploitation of the Past". Retrieved from https://www.vice.com/en/article/modern-warfare-the-highway-of-death-and-call-of-dutys-exploitation-of-the-past/.

Genette, G. (1997): Paratexts: Thresholds of Interpretation (J. E. Lewin, Translator). Cambridge: Cambridge University Press.

Gray, J. (2010): Show Sold Separately: Promos, Spoilers, and Other Media Paratexts. New York: New York University Press.

Hazan, S. (2020). "Deep Fake and Cultural Truth: Custodians of Cultural Heritage in the Age of a Digital Reproduction." In: Rauterberg, M. (ed.) Culture and Computing, 8th International Conference Proceedings. Cham, Switzerland: Springer Nature, pp. 65–80.

Heath, S. (1977): "Screen Images – Film Memory." Cine-Tracts 1(1), pp. 27–37.

Holsti, O. R. (2011): American Public Opinion on the Iraq War. Ann Arbor: University of Michigan Press.

IGN. (29 October 2012): "'Surprise' - Black Ops 2 Official Live-Action Trailer" [Video file]. Retrieved from https://www.youtube.com/watch?v=pyjicUdj59w.

IGN. (20 October 2016): "Red Dead Redemption 2 - Trailer # 1" [Video file]. Retrieved from https://www.youtube.com/watch?v=-4JHq7KLWZ4.

IGN. (27 May 2024): "Call of Duty: Black Ops 6 - Official 'Open Your Eyes' Teaser Trailer" [Video file]. Retrieved from https://www.youtube.com/watch?v=KBoeF8otBCE.

Infinity Ward (2007): Call of Duty: Modern Warfare [Video game]. Activision.

Infinity Ward (2009): Call of Duty: Modern Warfare 2 [Video game]. Activision.

Infinity Ward (2013): Call of Duty: Ghosts [Video game]. Activision.

Infinity Ward (2016): Call of Duty: Infinite Warfare [Video game]. Activision.

Lee, C. H. J. (22 July 2015): Sizzler and the Search for the American Dream. https://www.eater.com/2015/7/22/9011329/life-in-chains-sizzler-korea.

Lee, T./Hosam, C. (2020): "Fake News is Real: The Significance and Source of Disbelief in Mainstream Media in Trump's America." Sociological Forum 35(51), pp. 996–1018.

Makai, P. K. (2018): "Video Games as Objects and Vehicles of Nostalgia." Humanities 7(4), pp. 1–14.

Marchegiani, C./Phau, I. (2010): "Away from 'Unified Nostalgia': Conceptual Differences of Personal and Historical Nostalgia Appeals in Advertising." Journal of Promotion Management 16(1-2), pp. 80–95.

Mauran, C. (31 January 2025): "Thousands of Datasets From data.gov Have Disappeared Since Trump's Inauguration. What's Going On?" Retrieved from https://mashable.com/article/government-datasets-disappear-since-trump-inauguration.

McLaughlin, E. C. (29 December 2020): "How Atlanta Rappers Helped Flip the White House". Retrieved from https://www.cnn.com/2020/12/29/politics/atlanta-hip-hop-flip-election-senate-white-house.

McNicholas, C./Poydock, M. (8 May 2025): "Corruption in Plain Sight: How Elon Musk Has Benefited from the First 100 Days of the Trump Administration". Retrieved from https://www.epi.org/blog/corruption-in-plain-sight-how-elon-musk-has-benefited-from-the-first-100-days-of-the-trump-administration/.

Muehling, D. D./Pascal, V. J. (2011): "An Empirical Investigation of the Differential Effects of Personal, Historical, and Non-nostalgic Advertising on Consumer Responses." Journal of Advertising 40, pp. 107–122.

Muphy, G./Flynn, E. (2022): "Deepfake False Memories." In: Wang, Q. (ed.) Memory Online. London: Routledge, pp. 112–124.

Office of the Historian. (n.d.): "The First Gulf War". Retrieved 15 May 2025, from https://history.state.gov/departmenthistory/short-history/firstgulf.

Payne, M. T. (2012): "Marketing Military Realism in Call of Duty 4: Modern Warfare." Games and Culture 7(4), pp. 305–327.

Piero, M. (2024): "A Far Cry from Greatness: Christian Fundamentalism, Liberty Porn, and the 'Good People' of Trump's American in Far Cry 5." European Journal of American Culture 43, pp. 109–125. Retrieved from https://doi.org/10.1386/ejac_00115_1.

Rabin, N. (25 January 2006): "Stephen Colbert". Retrieved from https://www.avclub.com/stephen-colbert-1798208958.

Ramsey, D. X. (8 July 2020): "The Political Education of Killer Mike". Retrieved from https://www.gq.com/story/killer-mike-the-atlanta-way.

Redhouse, R. (3 September 2012): "Sizzler Promotional Commercial 1991" [Video file]. Retrieved from https://www.youtube.com/watch?v=E3YGtQ4oQvs.

Reuters (11 February 2025): "As Trump Hits Delete, the Race is on to Save LGBTQ and Climate Data". Retrieved from https://www.nbcnews.com/nbc-out/out-politics-and-policy/trump-hits-delete-race-lgbtq-climate-data-rcna191652.

Retropond (2 March 2021): "CHIA PET - 1990s Commercial" [Video file]. Retrieved from https://www.youtube.com/watch?v=GDatetEe9Os.

Rockstar Games [@RockstarGames]. (16 October 2016): "RED DEAD REDEMPTION 2 Coming Fall 2017 #RDR2". Retrieved from https://x.com/RockstarGames/status/788363842329903104.

Rockstar Games (2018): Red Dead Redemption 2 [Video game]. Rockstar Games.

Scharrer, E. (2004): "Virtual Violence: Gender and Aggression in Video Game Advertisements." Mass Communication and Society 7(4), pp. 393–412.

Seiwald, R./Vollans, E. (2023): "Introduction: Video Games as Networked Texts." In: R. Seiwald/E. Vollans (eds.), (Not) in the Game: History, Paratexts, and Games. Berlin/Boston: De Gruyter, pp. 1–11.

Seiwald, R. (2021): "The Ludic Nature of Paratexts: Playful Material in and Beyond Video Games." In: B. Beil/G. S. Freyermuth/H. C. Schmidt (Eds.), Paratextualizing Games: Investigations on the Paraphernalia and Peripheries of Play. Bielefeld: transcript, pp. 293–317.

Shapiro, A. (19 April 2010): "Distrusting Government: As American as Apple Pie". Retrieved from https://www.npr.org/2010/04/19/126028106/distrusting-government-as-american-as-apple-pie.

Singer, E. (2 February 2025): "Thousands of U.S. Government Web Pages have been Taken Down Since Friday". Retrieved from https://www.nytimes.com/2025/02/02/upshot/trump-government-websites-missing-pages.html.

Sledgehammer Games (2023): Call of Duty: Modern Warfare III [Video game]. Activision.

Stern, B. B. (1992): "Historical and Personal Nostalgia in Advertising Text: The Din de Siècle Effect." Journal of Advertising 21(4), pp. 11–22.

Švelch, J. (2015): "Towards a Typology of Video Game Trailers: Between the Ludic and the Cinematic." GAME The Italian Journal of Game Studies (4), pp. 17–21.

Tariq, S./Abuadbba, A./Moore, K. (2023): "Deepfake in the Metaverse: Security Implications for Virtual Gaming, Meetings, and Offices." WDC '23: Proceedings of the 2nd Workshop on Security Implications of Deepfakes and Cheapfakes, pp. 16–19.

Treyarch (2010): Call of Duty: Black Ops [Video game]. Activision.

Treyarch (2012): Call of Duty: Black Ops II [Video game]. Activision.

Treyarch (2015): Call of Duty: Black Ops III [Video game]. Activision.

Treyarch/Raven Software (2020): Call of Duty: Black Ops Cold War [Video game]. Activision.
Treyarch/Raven Software (2024): Call of Duty: Black Ops 6 [Video game]. Activision.
Ubisoft (2018): Far Cry 5 [Video Game]. Ubisoft.
Villegas, E./Fonts, E./ Fernández, M./ Fernández-Guinea, S. (2023): "Visual Attention and Emotion Analysis Based on Qualitative Assessment and Eyetracking Metrics: The Perception of a Video Game Trailer." Sensors 23(23), pp. 1–28.
Walters, J. (28 May 2025): "Denied, Detained, Deported: The Faces of Trump's Immigration Crackdown". Retrieved from https://www.theguardian.com/us-news/2025/apr/28/trump-immigration-people-detained-deported-cases.
Westerlund, M. (2019): "The Emergence of Deepfake Technology: A Review." Technology Innovation Management Review 9(11), pp. 40–53.
Wright, E. (2023): "Paratexts, 'Authenticity', and the Margins of Digital (Game) History." In: R. Seiwald/E. Vollans (eds.), (Not) in the Game: History, Paratexts, and Games. Berlin/Boston: De Gruyter, pp. 33–53.
Wright, E. (2018): "On the Promotional Context of Historical Video Games." Rethinking History 4, pp. 1–11.
Vollans, E. (2023): "Artefact, Advert, or Advertising? Getting to Grips with Game Trailers." In: R. Seiwald/E. Vollans (eds.), (Not) in the Game: History, Paratexts, and Games. Berlin/Boston: De Gruyter, pp. 161–175.
Yourish, K./Daniel, A./Data, S./White, I./Gamio, L. (7 March 2025): "These Words are Disappearing in the New Trump Administration". Retrieved from https://www.nytimes.com/interactive/2025/03/07/us/trump-federal-agencies-websites-words-dei.html.
Zimmer, B. (2005): "Truthiness or Trustiness". Retrieved from http://itre.cis.upenn.edu/myl/languagelog/archives/002586.html.

Queering Hong Kong
Modded History in *A Summer's End: Hong Kong 1986*

Diego Barroso Sánchez

Abstract

Hong Kong has seen its fair share of portrayals in video games (over 300) – some, where it plays a more central role than others; however, academia has shown little interest in exploring how the city's history and people are represented in video games. While there is historiographical value in how the city is portrayed in non-historical games (ludoforming and modelling are just as important), a more direct approach to how history is recorded gives us a different understanding of its development and potential future. This paper argues that video games offer affordances that allow designers and players to reflect and reassess history by 'modding' (modification in video game terms).

Movie-based videogames (like those based on the performances of Jackie Chan during the 70s and 80s, for instance) were for many years our only window to the colonial era in this medium; therefore, the recent release of interactive digital narrative game A Summer's End: Hong Kong 1986 (ASE) (Oracle & Bone 2020), which brings a momentous period in Hong Kong's colonial history to the fore, is of capital importance. Both the city and its main characters, Michelle and Sam (typical Hongkongers), find themselves steeped in uncertainty. First, by the coming Handover (1997), which signalled the end of British colonial rule of the city; and second, by their budding non-heterosexual relationship jeopardising Michelle's 'fairy tale' upbringing.

This paper offers an in-depth playthrough of ASE informed by Harmut Koenitz's hermeneutical approach to interactive digital narrative, supported throughout by an interview with the game's developers, where the fascinating implications of ludonarrative feedback from the perspectives of design and gameplay are discussed. Ultimately, this game is shown to be an example of modded history in several ways: not only does it portray the homosexual relationship of two young women in a light-hearted/heartening way, but this candidness also echoes a sense of optimism that pervades the hopeful ending of the game. Michelle and Sam can prevail as a couple just as Hong Kong can overcome the historical challenge that was/is the '1997 Handover'.

Keywords

Colonial/postcolonial Hong Kong, interactive digital narrative, queer game studies, video game historiography

1. Introduction: Historying through play

Video games relate to our historical reality in diverse ways (Chapman 2016: 18), even a game's genre and its specific features can shape our disposition (embodied and otherwise) and our engagement with historical discourses (ibid: 275). In the case of Hong Kong, what can these historical relations be, given its colonial past and postcolonial present? And what kinds of gameplay systems/narrative structures can be implemented to meet those expectations?

The history of Hong Kong is what we would like to describe as the unfolding of a 'transferred colonialism', in the sense that, unlike territories in Asia such as India and the Philippines that, after sustained conflict, liberated themselves from a colonising entity, the city itself returned (or will return from the game's time frame) to an ostensibly previous state, one of belonging to a bigger entity than itself, belonging to China[1]. A lack of direction, a feeling of being simultaneously an urban entity on its own and part of the Mainland, is a feeling that arises in Hong Kong and that is repeated in different ways and is visible in several aspects of the city's identity. In her exploration of queer culture in Hong Kong, Tang (2012) reminds us: "Predictions of a doomed culture, a failing society, and a place without hope have shaped public discussions and media representations of Hong Kong and influenced academic and literary circles." (597)

If we focus on the LGBTQ+[2] and their place in the sociocultural landscape of Hong Kong, we also find histories of struggle, constant persecution, and resistance; again, we see the aforementioned prevailing uncertainty. From the disgraceful prosecution of the British police agent John MacLennan in 1980 (cf. Collett 2020) to the troubled organisation of the Gay Games of 2023 (cf. GGHK 2023), Hong Kong has come a long way in terms of its queer presence and representation. However, despite the adoption of "anti-discrimination ordinances [that] prohibit direct and indirect discrimination, harassment, victimization, and vilification" (Barrow 2020: 149), there is still resistance and unwillingness to adopt or adapt to international human rights standards in matters of sexuality and gender recognition[3].

What playful experiences and narratives can arise or have arisen between the budding self-affirmation of the queer communities in Hong Kong and the deeply

---

1   The idea that Hong Kong is still in a state of colonialism, now under China, is an attractive notion to explain its current state; however, as Keung (2003) states: "Colonialism often starts with the assumption of racial inequality," (88) and in this case, the perceived ethnic inequality would in fact favor the Hongkongese.
2   We shall stick to the term 'LGBTQ+', which has been used at least until 2024 in reputable scholarly sources (cf. Hatzenbuehler et al. 2024). We use it interchangeably with 'queer' as we consider both to be as inclusive as possible.
3   For an interesting exploration on the 'vernacularisation' of human rights in Hong Kong, see Madson (2021).

entrenched sexual discrimination that still haunts the city? *A Summer's End: Hong Kong 1986* (ASE) (Oracle & Bone 2020) stands as a recent example of the way games and gameplay can establish a dialogue with the historical narrative of a city while making a distinct interpretation/commentary of the lived experiences of its population (with a focus on the LGBTQ+). In this paper, which moves beyond a close reading of the game through a historical/LGBTQ+ lens, we discuss the design process that led to the construction of a history-based narrative of Hong Kong in their first video game with the development team at Oracle & Bone. By going back and forth between historical experience/gameplay experience and exploration of the design process, we analyse the way new media has the potential to rethink historiography and how, as a consequence, our engagement with history can be transformed, too. Thus, we consider Oracle & Bone's maiden work about 1980s Hong Kong a historical game in many ways, an artefact "affording narrative history*ing*" (Chapman 2016: 193) that focuses on an intimate story consciously and deeply embedded in a specific historical context.

Harmut Koenitz's (2023) analysis and codifying of the genre he calls 'interactive digital narrative' (IDN) sheds light upon this medium's challenges in its use and existence. The attention he pays to the added levels of interactivity video games afford resonates with our focus on experiencing/writing history (history*ing*) in a twofold way: notionally, as we revise Hong Kong's history, and ludically, as we replay the game in search of counter-narratives or modded interpretations of the city on the eve of the 1997 Handover (Oracle & Bone n.d.). We set out to discuss "the broader possibilities and limitations of games to represent various historical cultures given that the play element of these cultures may have differed (or is claimed to have differed) over time" (Chapman 2019: 134), then we ground these finding within the widely accepted view that interactive digital narratives (or visual novels) constitute an especially strong medium for rewriting and rethinking established historical narratives (Mariani et al. 2023: 6).

Considering the historically conscious approach to the design of ASE, we are led to consider it as a work of historiography. Though the specific nature of its discourse may be defined by the facts known or available to the designers, we have decided to explore the events and dynamics that shaped Oracle & Bone Studio's attitude towards history, together with the intimate (hi)story surrounding the sexual/historical awakening that they crafted. As such, we look at the 1984 signing of the Sino-British Joint Declaration in tandem with sociopolitical unrest as providing a space for debate on queer identities to come to the fore.

The forums for discussion that appeared as the Handover approached have led to a permanent and generalized state of uncertainty for the identity of Hong Kong and a constant evolution of the LGBTQ+ presence in the city. The different forms of instability and wavering in Hong Kong do not necessarily translate into constructive developments; Yue and Leung (2017: 761) point out that cities like Singapore and Hong Kong manifest a form of "disjunctive logics", a lack of alignment between legislations, economic interests, cultural activities, and activism, which result in a contradictory heteronormativity. This reality imbues works such as ASE with a sense of a "queer utopia" (Muñoz 2009: 1), where reality is modded to accommodate non-heteronormative identities.

When we align the three elements – Hong Kong, video game (historical) representation, and queerness/queer identity – we find conceptual and methodological affinities that echo and harmonise through our analysis of the historical discourse articulated in ASE. Hong Kong has already been characterised as an uncentred, unsystematic developed city (at least the Kowloon Walled City; cf. AlSuwaidi et al. 2024), a rhizome extending with unexpected and exciting developments, and our reading of this virtual version of the city follows this decentralised perspective, from the point of view of (embedded) historiography.

2. Interactive Digital Narrative (games) as historiography, as modding

2.1 Video games and historiographic narrative

'Historiography' has already been considered a key lens for reading and understanding the spatial design of video games (for a study focusing on Hong Kong, see Barroso and Holopainen 2024); we also think of the medium as an important source of historical discourse, in this case for the narrative design of a game.

As Adam Chapman, one of the main scholars that focus on history in games (or games as history), says, when "the historical becomes that which attempts to make meaning out of the past" (ibid.: 11), the images and narratives from video games can be doubly valuable, both as a product of a specific time period and as a way to preserve/explore our personal and collective memories.

Several studies have explored the presence and uses of history in different media (for a general outline, see de Groot 2009); however, there is still a gap in the study of visual narratives in print (for works discussing history in comics, see Babic 2014; McKinney 2008) and digital platforms, including visual novels (or interactive digital narratives in general) as producers or carriers of historical narratives. The lack of attention given to video games with historical content/settings has been palliated in recent years with studies focusing on the historiography and historical affordances embedded in these games (such as Champion 2011; Fogu 2009; Stirling/Wood 2021), as well as more general explorations of history in games and historical games (Kapell/Elliott 2013; von Lünen et al. 2019).

Chapman (2016: 18) considers games to be an important form of historical narrative, for they not only allow the traditional reading of historical discourse, but they can also be "systems for *doing* history (historying)" (Ibid: 32, emphasis in the original). While we have given a fair share of importance to the *content* of ASE (as in its historical narrative) due to the intermedial status of interactive digital narratives, how this content is conveyed cannot be dissociated from gameplay.

We will pay greater attention to the affordances of the genre of interactive digital narratives, where choosing is doing, following Koenitz's in-depth analysis.

By engaging in a review of studies on the history of Hong Kong, queer representation/presence in media, and queer game studies, we set out to explore the city's recent history (from the decade of 1980) through the lens of interactive digital narrative and the experience of the double interaction of "interpretation plus impact on the system" in games (Koenitz 2023: 3), taking ASE as our focus. In this playthrough, we look at: (1) the historical events alluded to and discussed directly; (2) the developing queer identity of the protagonists; (3) the social climate refracted through both the aforementioned historical juncture, and the situation of the queer population (more specifically lesbian) in Hong Kong.

This analysis has been thoroughly enriched and contextualised by an interview/play session with Charissa So and Tida Kietsungden[4], founders of Oracle & Bone Studio and designers of ASE, the fundamental resource that brings us closer to game production studies (Engström 2019: 9; O'Donnell 2021: 151). In this dialogue, we went deeper into the design process that led to a reinterpretation of Hong Kong in times of the Handover in 1997, through the lens of the queer experiences of the main characters.

Even though this research aims to explore the possibilities of exploration and reassessment of historical discourses, we consider our work to be orthogonal to the field of Game Production Studies, a more circumscribed area of Game Studies (Sotamaa/Švelch 2021: 9). We are not tracing the whole production process, but how specific affordances made for historiographical aspects in Oracle & Bone Studio's debut work were conceived and executed. However, it is worth noting that both Charissa So and Tida Kietsungden fit in the schema that avoids what Brendan Keogh (2021) calls the "creativity dispositif" (34), the production system that leads game creators to justify a self-exploitation that legitimises them as full-time developers; for So and Kietsungden, their studio remains uncompromised as they each keep their previous occupations separated from the Studio[5].

2.2 Interactive digital narrative in depth

In the words of its designers, ASE is considered a visual novel. This type of game falls under the more general genre that Helmut Koenitz (2023) has called 'interactive digital narratives' (IDN) for its ability to incorporate specific attitudes and responses. He defines IDN as:

---

4   So, C./Kietsungden, T. (2024): "Interview by Diego Barroso Sánchez." (December 2024)
5   However, Tida Kietsungden's work beyond Oracle & Bone Studio has been design and/or game-design-related for a long time, even before the establishment of the company (cf. So/Kietsungden, interview).

[A] narrative expression in various forms, implemented as a multimodal computational system with optional analog elements and experienced through a participatory process in which interactors have a non-trivial influence on progress, perspective, content, and/or outcome where narrative is understood as a flexible cognitive frame for mentally projected worlds (Koenitz 2023: 5).

There are many elements to unpack in this definition. First and foremost, its 'narrative configuration' sets a foundation for historiography since "narrations outside traditionally printed books can be analysed more appropriately, on a cognitivist basis, as historiographic narrations" (Fulda 2014: 235). But besides the fact that the main medium for IDNs is a computational system, our focus is the involvement (or a form of 'nontrivial effort'; cf. Aarseth 1997) of the interactor (player) for the narrative to move forward, to come to fruition. Koenitz (2023: 93) further emphasises how the combination of agency with the consciousness of choice affords narrative experiences unique to the medium of video games.

One of the main features of the IDN is its double dynamic of interactivity (ibid.: 3); this idea refers to the fact that these works not only allow the player to speculate about the development of the plot but also to have an impact on the system. In this regard, the player's role is emphasised as a form of co-authoring, and this new form of narrative helps us "overcome this deeply rooted but limited comprehension of narrative as Eurocentric, non-interactive, and literary form mostly created by white men" (ibid.: 9).

Furthermore, we share the conviction of Mariani et al. (2023) that IDN affords a space for exploration, for "challenging dominant narratives and promoting inclusive storytelling", a space to reassess different forms of "hegemonic narrative" (ibid.: 5). They can even be considered an "attempt to reach a (at least assumed) personalized and interactive storytelling experience" (ibid.: 9).

2.3 The (inherent) possibilities of queering history through games

One of the key ideas that guided our reading of the game's historical narrative was José Esteban Muñoz's (2009) idea of queerness as future/hope/utopia: "Queerness is essentially about the rejection of a here and now and an insistence on potentiality or concrete possibility for another world." (1) This position made waves in the field of queer game studies: Macklin (2017) observed that a "queer world building" (251) in games helped players access the queer utopias ideated by Muñoz, while Harper et al. (2018) considered queerness "as a realm of possibility" (4). More recently, again referring to this utopic queer future, Belmonte Ávila and Encarnación-Pinedo (2024) analysed queer time in video games as not limited to the present of the game, but representing a "future-and-hope-bound queer horizon" (6–7).

We might say that queer games (more easily) afford a return to play (Ruberg 2018: 552), either by directly enacting a queer narrative or by *playing queer*, that is, counter-teleologically, challenging the game system and disrupting the hegemony implied by the (game) rules. And as history playing (another type of historying), ASE taps into these new ways of constructing a historical narrative: "Historical knowledge has its own performativity, and maybe it's possible to play with that creatively, to embrace the fluidity of our role in its constant social construction." (Street 2017: 41) As we shall see, the branching and diverging paths in ASE correspond with the notions of "potentiality" and "possibility" posed by Muñoz (2009), offering the player the ability to *mod* the history of the protagonists and their views of Hong Kong.

2.4 Modding our understanding of history

In video game communities, the concept of *mod* and its associated practices carry both positive and negative connotations. Short for *modification*, it involves any form of non-remunerated alteration to a released game, distributed online through dedicated forums and communities. Mods can range from aesthetic changes and rebalancing of game mechanics to more complex expansions that effectively enlarge or transform the original game, from repurposing assets to creations that come across as completely new games (Unger 2012: 518; Welch 2018).

Can we consider the act of creating a completely original video game to be a modding practice of cultural conventions or historical discourses? In their metareview of studies about mods, Reisinho et al. (2023) concluded that:

Mods' ability to revitalise a commercially or technologically invalid game, using co-creation practices grounded in participatory culture and remix culture, attests to their ability to unite players from different socio-cultural backgrounds and counter political, social, and cultural discourses that may dominate digital games (Reisinho et al. 2023: 886).

ASE does this for Hong Kong. Not only does it breathe new life and interest into a rapidly changing city whose media presence almost feels like a long-gone memory, but it also explores alternative narratives to the 'dominant' historical accounts of colonial Hong Kong, further emphasising the potentiality of queer identity and content. Then, the creators behind Oracle & Bone Studio become 'queer modders' whose work, in the words of Tom Welch, "become a powerful tool for destabilizing the accepted norms of both the video game industry and naturalised gender and sexuality performance" (2018 n. p.), though, in their case, the game being modded would actually be historiography.

3. A queer history of Hong Kong

For So and Kietsungden, ASE is a visual novel or dating sim (cf. Camingue et al. 2021). The embedded affordances of these genres might not be as involved as the 'emergent narrative' that Chris Crawford (game designer and Game Developers Conference founder) envisioned, where "the dynamic evolution of the story is based on user input and the system's responses, nurturing personalised and unique storytelling experiences" (Mariani et al. 2023: 8). Nevertheless, we find in the work of Oracle & Bone Studio a deliberate attempt to write a small and personal (hi)story of Hong Kong focusing on the underrepresented (lesbian relationships), even the often invisible (non-Han Chinese identities), strands of the city's social fabric.

Writing on the eve of Hong Kong's 1997 change of administration from a British colony to a Chinese Special Administrative Region, Ackbar Abbas (1997) wrote: "The anticipated end of Hong Kong as people knew it was the beginning of a profound concern with its historical and cultural specificity." (7) Abbas notices the formation of a postcolonial identity even before the end of Hong Kong's Handover (ibid.: 10); the creeping instability, a desire to define its own identity, resistance against disappearance, these are all fugues from "another stable identity" (ibid.: 15).

What Abbas sees in the budding interest in the way history relates to space, to Hong Kong, is the possibility to "modify some common assumptions about the forms that both colonialism and postcolonialism would take" (ibid.: 73). By engaging with Hong Kong's history, we notice the appearance and persistence of that 'concern' arising at a momentous crossroads for the city and how it permeated and gave way to other social dynamics or identities, often in contrast with Mainland sociopolitical tensions (Fung 2004: 401), but not always.

Let's look at the sociopolitical outlook leading to the time period modelled in the game (the early and mid-1980s), in a moment when a local identity started to form either as a "resistance against the presence of China" or as "accommodation during political transition" (Lo/Pang 2007: 402). We look at two strands of Hong Kong's history – the aftermath of the signing of the Sino-British Joint Declaration and the critical changes to the status of LGBTQ people in the city – that, in fact, are entwined in a string that resonates throughout the whole gameplay experience in ASE.

3.1 A shifting belonging: the Joint Declaration

When, in 1842, the British Empire drove its ships to Nanking (Nanjing) – too close to comfort for the Qing Empire (1636–1912) (Tsang 2004: 12) –, no one would have imagined that the island and the peninsula across its strait ceded to Queen Victoria (1819–1901) would grow into one of the trademarks global metropolises in the history of humanity in little more than a century. Oftentimes, the history of Hong Kong is written in contradictory ways, either *in medias res*, mainly when we look at earlier accounts from the West (even more recently, in 2007, Ingham addresses this issue while falling prey to it and dismissing the pre-colonial in a few pages), or back into a deep past as can be seen in more recent historiographical works from Mainland China (for a fascinating comparison of these historical narratives, see Yuen 2003).

Let us look at the events surrounding the signing of the Sino-British Joint Declaration in 1984, the definitive agreement that described the timing and format of the return of Hong Kong to Mainland China, with special attention to the effects this treaty had on the attitudes of the local population. The years leading to this bilateral agreement saw a changing attitude towards politics, which manifested itself as a lack of interest and involvement in general, and in practice as either an 'exodus' away from China or an open mockery of the Mainland (Chan 2022: 159). This was quite different from Tsang's (2004) description of the early 1980s, where "a distinct Hong Kong political culture had emerged and local expectations had changed. The people of Hong Kong by then increasingly wanted democracy" (206).

The Joint Declaration consisted of a roadmap for Hong Kong's future after the return of sovereignty to the PRC (a 'blueprint', calls it Tsang 2004: 227; a precursor for the Basic Law, according to Flowerdew 2012: 177). The Declaration would create a lot of expectations amongst the locals since with it, "Britain and China agreed to a gradual introduction of a limited number of seats to the Legislative Council" (Flowerdew 2012: 47). This outline was met with concern, as Chan (2022) reveals, and the matter of a "Hong Kong identity" (158) suddenly came to the fore. As the Handover drew nearer and nearer, events in the Mainland transformed uncertainty into fear, which in turn triggered an 'exodus' to countries across the world (Canada, for instance, being one of the preferred destinations; cf. Li 2005).

As mentioned above, the idea that such a territory, notwithstanding its disadvantageous location, could rise to the heights of global (mostly economic) development was elevated to mythical levels towards the end of its British era. As Chris Patten, the last British governor of Hong Kong, declared in a speech in 1994:

"Since 1984: living standards for the whole community have improved dramatically. Total GDP has grown by 79 percent in real terms, and in terms of GDP per person, we now rank 17th in the world instead of 28th, as we were in 1984; the value of our foreign trade has increased by about 350 percent in real terms." (quoted in Flowerdew 2012: 53)

Economic prosperity, flaunted as Hong Kong's success, conceals social tensions that come to be seen as obstacles to the former. As Tang (2011) reflects, a "regulatory discourse on economic progress" (15) helps understand the impact an official discourse has on the opening and closing of spaces for queer expressions.

Time after time, we are faced with a narrative written by authoritarian powers, or the white (British), or the male (heteronormative), the elite, a history that emphasises the economic; this position of the historical narrative conceals, to a lesser or greater degree, the presence and agency of other less dominant groups of actors in the history of Hong Kong. Lesbian, gay, and other non-cisgender identities have endured a history characterised by hardships and challenges, very much like the city itself. We shall briefly describe how the LGBTQ+ community found a way for their voices to be heard in the city's social and cultural spaces, before examining how these voices appear in ASE.

3.2 Queer in Hong Kong, from witch-hunt to frontal challenge

Rey Chow (1992: 157) describes Hongkongese identity as being "in-between" (again, an ambiguous state), as always being in "a particular kind of negotiation". It is no wonder that the identity/community-building process for the LGBTQ+ in the city parallels or resonates with the larger historical processes described in the previous section. As Helen Leung points out:

"The debates over the decriminalization of homosexuality throughout the 1980s resulted not only in the emergence of gay and lesbian identities and organized activism around these identities but also in a new discursive space where issues of sexual and gender transgressions can be openly voiced." (Leung 2008: 73)

We follow, then, Alvin K. Wong's guidance when he says that "[p]aying close attention to issues of feminism and queer desire can also unbind Hongkongness from the repetitive resentment and obsession of being "Chinese or 'not Chinese' enough" (2018: 1118).

A key event that brought LGBTQ+ identities to the fore in Hong Kong was the MacLennan Case (Tang 2011: 26). While homosexuality had already been vindicated to some degree in Britain (decriminalised in 1967), up until the 1980s the colonies still held on to a more conservative position when it came to same-sex relationships–particularly concerning sodomy (Han/Mahoney 2014: 2070). The MacLennan Case involved the Royal Hong Kong Police Force Inspector John MacLennan, who in 1980 was arrested "on eight counts of gross indecency" and his subsequent "suicide" before he was formally prosecuted (Johnson 1983: 73).

One of the main aggravators of this case came from the misunderstanding of homosexual relationships; at that point in time, there was a conflation of homosexuality with paedophilia, for it was assumed that gay men sought out younger or underage male company (Wesley-Smith 1982: 4). It appeared that the attention this case received, to quote Henry James Lethbridge, was tantamount to the "government opening 'Pandora's box'. It was opened, and one of the victims was the Commissioner of Police [Roy Henry, the individual in charge of dealing with MacLennan]" (1982: 29).

Nevertheless, in the long run, the effects of this case were not altogether negative. We have already said that the MacLennan case (together with smaller but similar ones that came before and after; for a full overview of the MacLennan case and others leading to it, see Collett 2020) motivated a series of well-publicised open discussions on the topic of homosexuality: "Although the media coverage on the MacLennan case and the issue of homosexuality has been negative yet it gave rise to a discourse on sexualities among lesbians, gays, and bisexuals in the community." (Tang 2011: 27)

The discourse and debates leading to and around the decriminalisation of homosexuality in Hong Kong (with the recommendation of a 1983 report from the Law Reform Commission of Hong Kong, implemented much later in 1991) were clearly focused on the legal implications of such actions. Still, as Leung (2008: 73) reflects, this opening space would quickly move onto the cultural sphere, mainly onto film. Open discussion of homosexuality would reach a milestone in 1989 with the appearance of the "tongzhi" identity (one of the main labels adopted by the LGBTQ+ to identify themselves, cf. Tang 2011).

Despite those openings and changes, LGBTQ+ characters in popular media were "made fun of, belittled or portrayed as deranged individuals", even at the turn of the century (Tang 2011: 41). A more recent survey of the "media and population in Hong Kong" (Yeo 2019: 251) confirms this negative attitude towards "sexual minorities", with the internet and social platforms offering a more open (and anonymous) space for gays, though less for lesbians (ibid: 253–4). While the Legislative Council has responded and modified the law in Hong Kong to combat discrimination, albeit slowly, its policies referring to sexual orientation and gender still lacked direct actions against implicit and explicit social attitudes; as Amy Barrow (2020) indicates in her study of anti-discrimination legislation in Hong Kong, "there remains deep-seated social conservatism around gender roles and gender-variant sexual identities" (127).

Denise Tang (2011) underlines the spatial restrictions that a city such as Hong Kong present for queer women (lesbian and transgender alike) but puts a positive spin on this challenge: "Hong Kong women and transgender persons are constantly seeking creative solutions and building capacity within their own networks to foster relationships with each other." (20) The creatives behind Oracle & Bone were all too aware of those 'new discursive spaces' that Leung identified at the beginning of the 1980s and the creative explorations that Tang found in the current urban ethos, and turned them into the leading motifs of their game. Charissa So and Tida Kietsungden's ASE also comes as a response to the lack of LGBTQ+ representation in "Asian media or in media in general, especially when it comes to lesbian relationships"[6].

4. Looking back 35 years through A Summer's End: Hong Kong 1986 with Oracle & Bone Studio

We have discussed how the genre of IDN helps us rethink our understanding of narrative; we would like to add historiography to the concept of 'narrative' (in some capacities, to some degree; cf. Dray 1971). This way, we may explore the historiographic possibilities of ASE in the realm of IDN; our brief overview of Hong

---

6 So/Kietsungden, interview.

Kong's queer situation around the 1980s has helped us reconsider Oracle & Bone Studio's work vis-à-vis mainstream historical discourse on the Handover years. In the case of ASE, the value of its historical narrative is double-layered, first as an implicit exposure to "history and historical representations" contained in the game (Chapman 2016: 14), and second as a historical self-consciousness during the game's development. Charissa So shared in our interview:

"[A Summer's End] is many things: it's an allegory, it's a 'coming-of-age', it's an LGBTQ story, but it's not just that, it's about identity, family, who you are in relation to your culture, your city…it tells a lot about not just the political conditions of the time then, but even now."[7]

Aligned with the view of IDN as a fertile ground for counter-narratives, queer theory and queer reading of history pull us toward the redistribution of meaning and the reinsertion of invisible subjects into historical consciousness. In her 'reparative reading' of ASE, Poirier-Poulin had already pointed toward the game's capacity for reassessing the past:

"In ASE, revisiting the past becomes a way to disrupt reality, to make history less limited, to stay hopeful, and to trust the future. At the end of the game, Sam is working on an archival film project and on the screening of queer films that have been overlooked; she thus seeks to reopen a past that was symbolically closed". (Poirier-Poulin 2022: 48)

We return to Mariani et al. (2023: 15) and the affordance for subversion, revision, and modding inherent to the interactive digital narrative genre. Not only do video games allow the player to challenge dominant narratives of history, but, more so than other formats, virtual novels have lent themselves to explorations that go beyond fiction, that try to reinsert excluded/silenced narratives and restructure well-entrenched ones; interestingly, in the case of Hong Kong in ASE, the dominant voices would be the majority Han Chinese of Guangdongese ethnicity (or even Shanghainese), over the also Han Chinese but of Hakka (Kejia) and Teochow (Chaoshan) ethnicity (Guldin 1984: 151), implying a fundamental separation between the Chinese and the non-Chinese, the latter of which are completely absent from the game[8].

---

7   So/Kietsungden, interview.
8   The white foreigners are generally referred to in Hong Kong as 'gweilo' (鬼佬), meaning 'ghost man'. While Guldin (1984) sees a greater distance between Han ethnic groups than that between the "local" ethnic Guandongese (142), other studies still find a boundary separating both the expatriate and the long-term foreign resident from the Chinese (vid. Leonard 2010). We won't explore further the latter boundary because it is not explored in ASE, but it is worth keeping it in mind when we look at portrayals of Hong Kong.

The next part of our study, its heart, we might say, blends our gameplay experience (our 'well-playing' as a fundamental approach to game analysis; cf. Bizzocchi/Tanenbaum 2011) with an in-depth semi-structured interview with Charissa So and Tida Kietsungden[9], the masterminds behind ASE. For our design, we followed Kallio and her colleagues' guide to structuring interviews (2016: 2962), skipping the "pilot testing" stage due to time and availability limitations.

Before, during, and after the interview, we identified certain themes that resonated with our prior analysis of ASE, and "moving back and forward between" our historical narratives, the game, and the interview (Braun/Clarke 2006: 86), we established three main elements that were key to writing a 'modded' historical narrative for Hong Kong: minority (sexual and ethnic), migration, pop culture (media). This reading is aligned with Donald and Reid's analytical framework for historical narratives in games, where the authenticity pursued by ASE is closely related to how its setting – and discourse – "'feels' in relation to personal, communal, and collective memory" (Donald/Reid 2023: 66).

4.1. Choosing present or presence: a queer utopia mod for Hong Kong's history

Oracle & Bone Studio's debut work was published on April 23, 2020, for Windows, Mac, and Linux computers (MobyGames 2023: Moby ID 144820). Soon after release, it was shown to resonate with the gaming community, which recognised in it a story unlike most in the medium. Bonnie Qu, reviewing the game for Polygon (2020: n. p.), appreciated that such a story could be told in her home city of Hong Kong: "it's incredibly gratifying to see the game establish that they [Sam and Michelle] don't have to compromise parts of themselves or leave Hong Kong to achieve the better future that everyone strives for".

The plot of the game itself is not complex in that it does not engage directly with events involving the destiny of the city (we have already stated that the developers view the game as very intimate and small-scale, using terms such as 'coming-of-age', 'family' etc.): the player follows Michelle, an office lady (she is even referred to as 'OL', a popular Japanese slang word widely used in Hong Kong)[10] who is making her way up the career ladder; a fortuitous event leads her to Samantha, a seemingly polar opposite: a video store owner who takes life as it comes, candid and sincere. Passion and confusion ensue; Michelle finds it difficult to understand her fascination with Samantha, who wants to stay close to Michelle, to really know and be with her. We will discuss specific moments in the game later

---

9   So/Kietsungden, interview.
10 There is a marked presence of Japanese popular culture. ASE has a few examples, such as references to products (like the mosquito repellent Mopiko) and pop idols (Michelle is compared to Momoko Kikuchi, who made her idol debut in 1984).

on, but we should always keep in mind that theirs is a relationship conscientiously situated in time and place.

One key element that shapes the narrative at large, and the unfolding relationship between Sam and Michelle in particular, is the aforementioned Joint Declaration. Set two years after the agreement was signed (1984), the character's actions and motivations are very much shaped by what that meant for the future of Hong Kong, and this holds true for non-playable characters (NPCs), too. For example, a promising male colleague of Michelle, Joey, was educated in the US and only came back to Hong Kong to find a suitable partner to bring back to America (Michelle is implied to be one such candidate):

Joey: You don't think about immigrating?
[...]
Joey: They signed the papers! You think things will be the same?

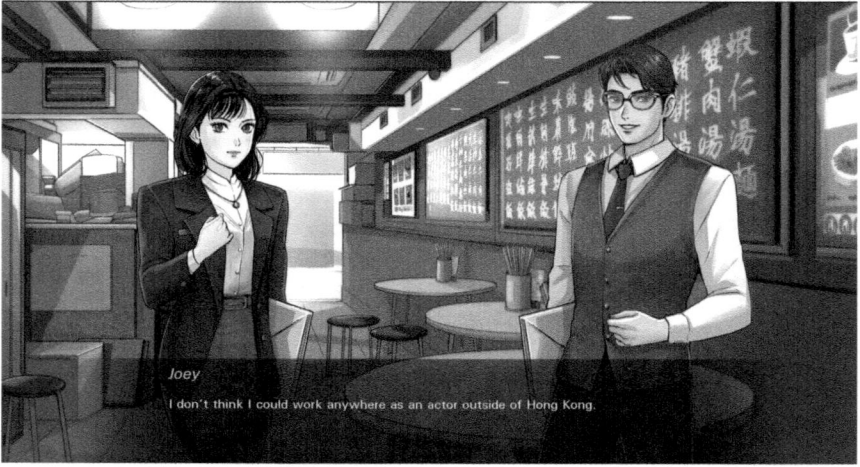

Fig. 1: Joey and Sam discuss immigration.

However, the game does not establish a teleology for the player (not even for Hong Kong); there is no implication of "right or wrong", but just "life", as Charissa So[11] declared. Despite the fatality that has led the city to its present situation, ASE presents a 'pivot' that turns with hope. Towards the end of the game, after Michelle and Sam finally share a night together, the former faces the dilemma of honouring her filial commitment to her mother or fully accepting the love she feels for Sam deep inside. Let us analyse this moment in-depth since it embodies and echoes a similar dilemma for Hong Kong.

---

11   So/Kietsungden, interview.

This aforementioned 'pivot' is procedurally (Bogost 2007) presented by an in-game choice that has a bearing on the ending sequence. After ignoring Sam (and her own feelings toward her), Michelle decides to have dinner with Joey, but by chance, they run into Sam, who hurries to leave. Joey asks if Michelle knows her (Sam). At this point, the player has a choice: the first choice has Michelle ignoring Sam and going through with dinner, a decision that leads to Michelle marrying and giving birth to a daughter, finally migrating to the US as the 1997 Handover takes place; the other choice, to follow Sam outside and give the relationship a chance, finds Michelle giving Hong Kong a chance too.

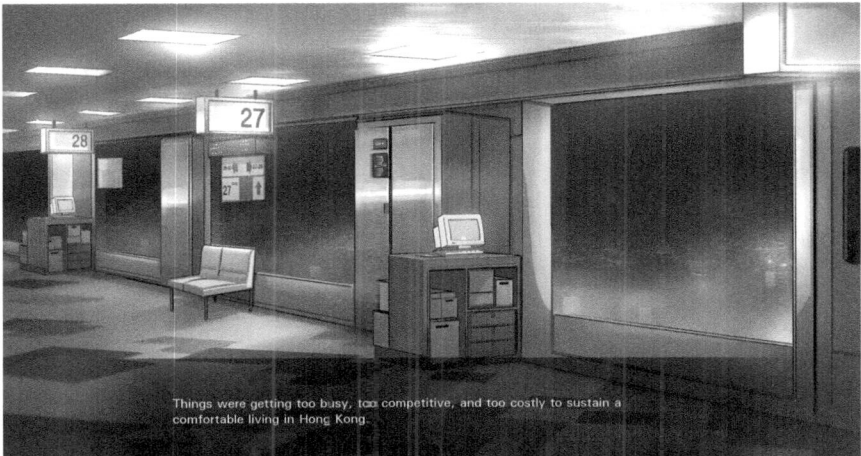

Fig. 2: Michelle is about to leave Hong Kong behind with her daughter.

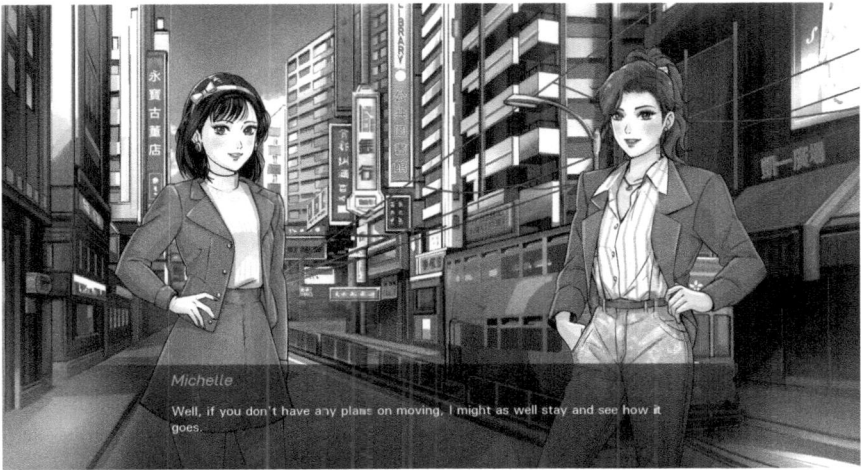

Fig. 3: Michelle gives the city a chance with Sam.

The second choice leads to a longer ending sequence–and more gameplay–that further explores these themes of uncertainty and hope. Michelle and Sam have an impromptu date in the city, with no particular destination; as the day starts to wind down, the couple gets on a *ding-ding train*, and as the conversation turns to travel abroad, Michelle asks Sam: "Do you ever think about where you would be after 1997?" Sam feels her life is in Hong Kong, and she wouldn't want to 'uproot' herself and start over elsewhere. Encapsulated in this brief dialogue, we find two opposing attitudes towards the deep ties the citizens may show towards the city, the place itself, and the life/ideals it represents. Michelle does not let up; she continues: "Do you really see yourself living freely then as you do now?" Does this uncertainty, this future projection of nostalgic sentiment, refer to Hong Kong, or her own previous (non-queer) identity? Coming full circle, back (forward?) to the optimistic futures/utopias, Samantha stands firm: "The future's not decided yet. We have the ability to decide it for ourselves." For Kietsungden, this decision is one and the same, no matter the ending: "Just like Michelle said in the end, we just want to do our best; we want our family."[12]

4.2. Between staying and leaving: rethinking migration and identity

Another aspect that is being rewritten or revised through ASE's narrative is that of migration. We have already mentioned Michelle's attitude towards the precarious situation of Hong Kong at that time and her wavering (sexual) identity echoing/being echoed by the prospect of leaving Hong Kong (with a male partner). In a study on attitudes towards migration after the handover, Salaff (2000) points out that, among several factors that influenced the decision to stay or leave Hong Kong, "(t)he longer [people] lived in Hong Kong, the more they were used to local lifestyles and felt they would not be able to adjust to life under mainland rule (264)." These findings align with Sam's attitude, who cannot imagine herself leaving her life in the city, and in turn, this thinking is reinforced by the opening of queer spaces in Hong Kong:

"The debates over the decriminalization of homosexuality throughout the 1980s resulted not only in the emergence of gay and lesbian identities and organized activism around these identities but also in a new discursive space where issues of sexual and gender transgressions can be openly voiced." (Leung 2008: 73)

Indeed, Michelle's sexual awakening and her coming to terms with her love for another woman are at the center of ASE; additionally, its exploration of other less visible identities in and of Hong Kong helps round off its historical narrative. Through the deuteragonist, Chinese ethnic minorities are reinserted into the

---

12 So/Kietsungden, interview.

history and fabric of the city: Michelle's father is Toishanese (from Taishan, Guangdong), and her mother is Teochow (from the region of Chaoshan), while Sam's mother was Hakka (also known as "Kejia", and distributed across different areas of Southern China). These groups are considered Chinese minorities. Still, the attitude of the Cantonese majority toward them and their languages can be negative/derogatory (Guldin 1984: 143) and has led to the erasure of these different denominations through "Guandongnese chauvinism" (ibid: 141) or "Han chauvinism" (ibid.: 146). As for the lesser-known parts of the Special Administrative Region (SAR) of Hong Kong, ASE reveals a side of the province that makes up most of its territory, but is hardly ever explored: nature[13]. By taking us to Sai Kung (a mere 30-minute ride from downtown Kowloon), the game helps us discover beaches in Hong Kong, ancestral villages that did not belong to the Han (majority), where it is possible to see starry skies close to one of the largest and brightest cities on Earth; once again, our understanding of Hong Kong is molded by our interaction with the game, and we find that this is a city that also has room for, as Tida said, slowing down and connecting intimately with others.

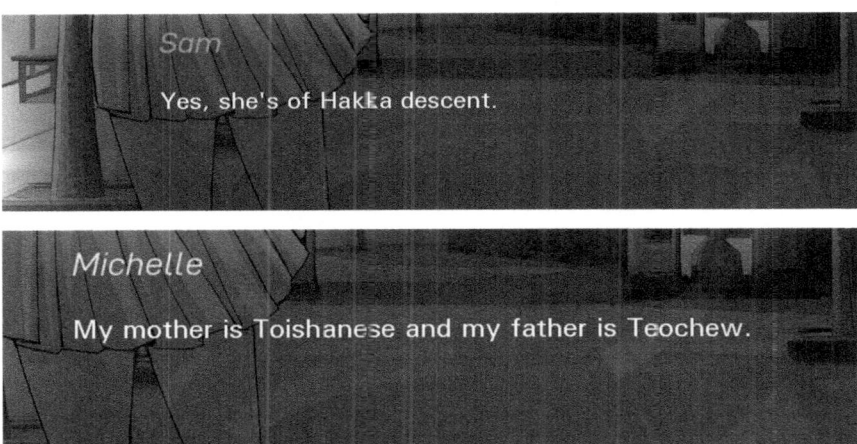

Fig. 4 and Fig. 5: Sam and Michelle share their ancestry.

---

13 There's much to say about nature in video games and the lack of scholarship around this subject until five or six years ago. Hong Kong's case is symptomatic, for less than 30% of its total area has been developed for habitation, and 40% is protected land (cf. The Nature Conservancy Hong Kong Foundation 2025). Yet, there is practically no representation in media, let alone in scholarly works.

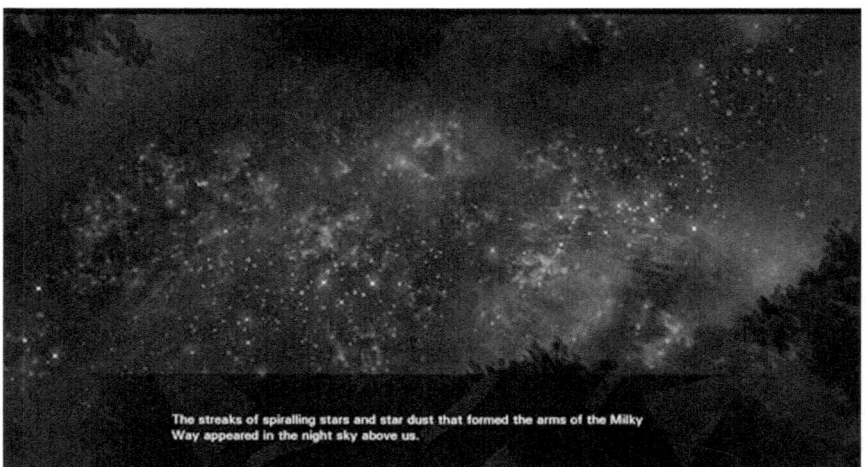

Fig. 6: *Starry sky in Hong Kong (Sai Kung)*.

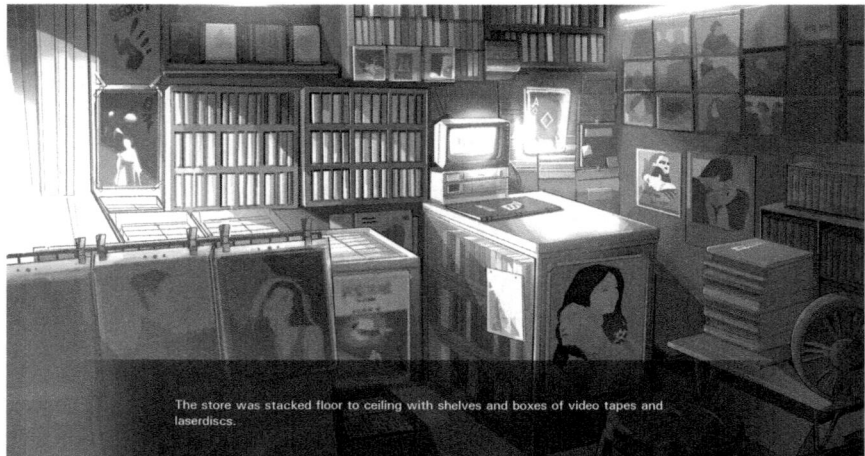

Fig. 7: *Sam's video store, to the brim with 'actual pop culture'*.

Besides the narrative, several elements that went into the game's design bear a deliberate purpose of history*ing* (Chapman 2016: 32) *this* Hong Kong, one that was and one that might have been (and might still be). As So indicates, Hong Kong itself is "a great setting to use"[14], however, in ASE, the city itself is transformed in consonance with the game's themes. Kietsungden reveals that the design of the game city was "romanticised"[15], polishing out some of the darkness usually associated with Hong Kong. Nevertheless, the stores, discos, movies, posters, and all these elements that may be lost to many players help build/rebuild this

---

14 So/Kietsungden, interview.
15 So/Kietsungden, interview.

"time period very much canonized in Hong Kong film and media"[16]. We may take these intertextual elements as paratexts, that is, "(t)he links between [...] in-game elements, the hardware, and the player" (Seiwald 2023: 18); from this perspective, ASE could even be considered a paratext of a new Hong Kong historiography, one that, according to Oracle & Bone Studio, encompasses its 'actual pop culture' – Hongkongese/Chinese and otherwise –, its LGBTQ people, its multiethnicity.

5. Conclusion: 'Timelessness' based on history

In one of the key works exploring video games as historiographic works, Matthew Kapell and Andrew Elliott (2013) warned that "when explaining or exploring the past, we are liable to impose upon those events modern values, meanings, and motivations in order to make sense of the actions of our distant forebears" (7). This threat looms over the way we build our historiographical discourses. During the interview, this very motif kept appearing: the situatedness/presentness of the game is informed by the designers' understanding of the current situation of Hong Kong, and this contemporaneity is then reflected in the portrayal of the past/history of the city: a game based on history, 'on actual pop culture', but also 'timeless'.

In this paper, we discussed the ability of interactive digital narrative games to explore and reassess history, which led us to consider them of historicgraphic value. As repeated throughout, this medium has offered a space for alternative narratives, both historical and not, and, as such, it is also a space for queer narratives and discourses to thrive. Taking advantage of those openings presented by video games, we set out to explore alternative or complementary views of Hong Kong's history, an important hub at the forefront of global economic exchange. Our gateway for this exploration was *A Summer's End: Hong Kong 1986*, a virtual novel that does not explicitly try to transgress or question the mainstream discourse on the history of Hong Kong, but that, nevertheless, manages to put a spotlight on less visible identities and ethnic groups.

The queer situation in the city, as we have mentioned, entered a crisis period due to internal changes and dissatisfaction, as attested by the unfolding of the MacLennan case, and by the sociopolitical changes looming with the Handover. Lucetta Kam (2017) observes the "general failure in Hong Kong people to connect gender/sexuality and political resistance" (167), when in fact queer action (or the manifestation of queer identity) in a city such as Hong Kong *is* political action. ASE brings this reality closer to the player: as the narrative unfolds, the player gets to know the social mores of Hongkongese (Chinese?) society; furthermore, they are faced with choices that might seem inimical to the fate of the city in the face

---

16 So/Kietsungden, interview.

of its return to Chinese sovereignty, but that are rhetorically parallel to the larger destiny of the local population.

Ultimately, ASE introduces a modded history in contradictory and complimentary ways: it brings to the fore the ever-evolving identity of the Hongkongese as it is borne out of political and social uncertainty; furthermore, it tries to turn visible the invisible, in this case, to give voice to communities that are either considered minorities in Hong Kong or are completely ignored beyond the city (e.g., ethnic groups such as the Hakka and Teochew, or queer sexual identities). The interview conducted with Charissa So and Tida Kietsungden (the former's family migrated to Canada from Hong Kong, and the latter migrated herself) about ASE and its development process reminds us that Hong Kong's identity has been shaped and is shaped by migration, and thus, it extends beyond borders, overseas, and across time.

References

Abbas, A. (1997): Hong Kong: Culture and the Politics of Disappearance. Hong Kong: Hong Kong University Press.

AlSuwaidi, M./Boussaa, D./Furlan, R./Awwaad, R. (2024): "The Paradox of Kowloon Walled City: Architectural Anomaly and Social Microcosm." Sustainability 16(15), 6515.

Babic, A. A. (ed.). (2014): Comics as History, Comics as Literature: Roles of the Comic Book in Scholarship, Society, and Entertainment. Lanham: Fairleigh Dickinson University Press/Rowman & Littlefield.

Barroso, D./Holopainen, J. (2024): "Play to Make a City. Cultural-historical Identities and Affordances in Videogame Versions of Hong Kong." Proceedings of DiGRA 2024 Conference: Playgrounds. Retrieved from: https://dl.digra.org/index.php/dl/article/view/2242/2239.

Bizzocchi, J./Tanenbaum, J. (2011): "Well Read: Applying Close Reading Techniques to Gameplay Experiences." In: D. Davidson (ed.), Well Played 3.0: Video Games, Value and Meaning. Pittsburgh: ETC Press, pp. 289–316.

Bogost, I. (2010): Persuasive Games: The Expressive Power of Videogames. Cambridge: MIT Press.

Braun, V./Clarke, V. (2006): "Using Thematic Analysis in Psychology." Qualitative Research in Psychology 3(2), pp. 77–101.

Champion, E. (2011): Playing with the Past. London: Springer London.

Chapman, A. (2016): Digital Games as History: How Videogames Represent the Past and Offer Access to Historical Practice. London: Routledge.

Chapman, A. (2019): "Playing Against the Past? Representing the Play Element of Historical Cultures in Video Games". In: A. von Lünen/K. J. Lewis/B. Litherland/P. Cullum (eds.), Historia Ludens: The Playing Historian. London: Routledge, pp. 133–154.

Chow, R. (1992): "Between Colonizers: Hong Kong's Postcolonial Self-Writing in the 1990s." Diaspora: A Journal of Transnational Studies 2(2), pp. 151–170.

Collett, N. (2020): A Death in Hong Kong: The MacLennan Case of 1980 and the Suppression of a Scandal. Hong Kong: City University of Hong Kong Press.

Crawford, C. (2005). Chris Crawford on Interactive Storytelling. Berkeley: New Riders.

Donald, I./Reid, A. (2023): "Account, Accuracy, and Authenticity: A Framework for Analysing Historical Narrative in Games". In: R. Seiwald/E. Vollans (eds.), (Not) In the Game: History, Paratexts, and Games. Boston: De Gruyter Oldenbourg, pp. 57–80.

Engström, H. (2019, January). "GDC vs. DiGRA: Gaps in Game Production Research." In: Proceedings of DiGRA 2019 Conference Game, Play and the Emerging Ludo-Mix. Retrieved from: https://dl.digra.org/index.php/dl/article/view/1081/1081.

Flowerdew, J. (2012): Critical Discourse Analysis in Historiography. The Case of Hong Kong's Evolving Identity. London: Palgrave Macmillan.

Fogu, C. (2009): "Digitalizing Historical Consciousness." History and Theory 48(2), pp. 103–121.

Fulda, D. (2014): "Historiographic Narration." In: P. Hühn/J. C. Meister/J. Pier/W. Schmid (eds.), Handbook of Narratology. Berlin: Walter de Gruyter, pp. 227–240.

Fung, A. (2004). "Postcolonial Hong Kong Identity: Hybridising the Local and the National." Social Identities 10(3), pp. 399–414.

"Gay Games 11 Hong Kong 2023" (GGHK) (2023): Retrieved from: https://www.gghk2023.com/.

Guldin, G. E. (1984): "Seven-veiled Ethnicity: A Hong Kong Chinese Folk Model." Journal of Chinese Studies 1(2), pp. 139–156.

Han, E./O'Mahoney, J. (2014): "British Colonialism and the Criminalization of Homosexuality." Cambridge Review of International Affairs 27(2), pp. 268–288.

Harper, T./Taylor, N./Adams, M. B. (2018): "Queer Game Studies: Young but not New". In: T. Harper/M. B. Adams/N. Taylor (eds.), Queerness in Play. Cham: Springer International Publishing, pp. 1–13.

Hatzenbuehler, M. L./Lattanner, M. R./McKetta, S./Pachankis, J. E. (2024). "Structural Stigma and LGBTQ+ Health: A Narrative Review of Quantitative Studies." The Lancet Public Health 9(2), pp. e109–e127.

Ingham, M. (2007): Hong Kong: A Cultural History. New York: Oxford University Press.

Johnson, H. (1983): "Scandal in Hong Kong." The Medico-Legal Journal 51(2), pp. 70–84.

Kallio, H./Pietilä, A. M./Johnson, M./Kangasniemi, M. (2016): "Systematic Methodological Review: Developing a Framework for a Qualitative Semi-structured Interview Guide." Journal of Advanced Nursing 72(12), pp. 2954–2965.

Kam, L. Y. (2017): "Return, Come Out: Queer Lives in Postcolonial Hong Kong." In: Y. W. Chu (ed.), Hong Kong Culture and Society in the New Millennium: Hong Kong as Method. Singapore: Springer, pp. 165–178.

Kapell, M. W./Elliott, A. B. (2013): Playing with the Past: Digital Games and the Simulation of History. New York: Bloomsbury Publishing USA.

Keogh, B. (2021): "Hobbyist Game Making Between Self-Exploitation and Self-Emancipation." In: O. Sotamaa/J. Švelch (eds.), Game Production Studies. Amsterdam: Amsterdam University Press, pp. 29–45.

Koenitz, H. (2023): Understanding Interactive Digital Narrative: Immersive Expressions for a Complex Time. London: Routledge.

Leonard, P. (2010): "Organizing Whiteness: Gender, Nationality and Subjectivity in Postcolonial Hong Kong." Gender, Work & Organization 17(3), pp. 340–358.

Lethbridge, H. (1982): "Pandora's Box: The Inspector MacLennan Enigma." Hong Kong Law Journal 12(1), pp. 4–30.

Leung, H. H. S. (2008): Undercurrents: Queer Culture and Postcolonial Hong Kong. Vancouver: UBC Press.

Li, P. S. (2005): "The Rise and Fall of Chinese Immigration to Canada: Newcomers from Hong Kong Special Administrative Region of China and Mainland China, 1980–2002." International Migration 43(3), pp. 9–34.

Lo, K. C./Pang, L. (2007): "Hong Kong: Ten Years After Colonialism." Postcolonial Studies 10(4), pp. 349–356.

Macklin, C. (2017): "Finding the Queerness in Games." In: B. Ruberg./A. Shaw, Queer Game Studies. Minneapolis: University of Minnesota Press, pp. 249–257.

Mariani, I./Ciancia, M./Ackermann, J. (2023): "Interactive Digital Narratives and Counter-Narratives. Systematising Knowledge to Derive Clusters as Lenses of Observation." GAME Journal 11(1), pp. 5–32.

McKinney, M. (2008): "Representations of History and Politics in French-Language Comics and Graphic Novels." In: M. McKinney (ed.), History and Politics in French-language Comics and Graphic Novels. Jackson: University of Mississippi Press, pp. 3–24.

Muñoz, J. E. (2009): Cruising Utopia: The Then and There of Queer Futurity. New York: New York University Press.

O'Donnell, C. (2020): "Game Production Studies: Studio Studies Theory, Method, and Practice." In: P. Ruffino (ed.), Independent Videogames. New York: Routledge, pp. 148-159.

Oracle & Bone (2020): A Summer's End: Hong Kong 1986 [Video game]. Oracle & Bone Studio.

Oracle & Bone (n. d.): "Hong Kong, Hong Kong". A Summer's End. Retrieved from: https://www.asummersend.com/blog/hong-kong.

Poulin-Poirier, S. (2023): "Trust, Confidence, and Hope in A Summer's End – Hong Kong 1986: A Reparative Reading." GAME Journal 11(1), pp. 33–52.

Qu, B. (2020): "A Summer's End is a Rare Queer Romance that Goes Where Most Games Won't". In: Polygon, 30 June. Retrieved from https://www.polygon.com/2020/6/30/21304214/summers-end-hong-kong-1986-love-visual-novel-romance-hope.

Reisinho, P./Raposo, R./Zagalo, N. (2024): "The Ethical Dilemma of Modding Digital Games: A Literature Review of the Creation and Distribution of Mods." Convergence 30(2), pp. 860–881.

Ruberg, B. (2018): "Queerness and Video Games: Queer Game Studies and New Perspectives Through Play." GLQ: A Journal of Lesbian and Gay Studies 24(4), pp. 543–555. Available at: https://read.dukeupress.edu/glq/article-abstract/24/4/543/135969/Queerness-and-Video-GamesQueer-Game-Studies-and.

So, C./Kietsungden, T. (2024): "Personal Interview," December 2024.

Sotamaa, O./Švelch, J. (2021): "Introduction: Why Game Production Matters." In: O. Sotamaa/J. Švelch (eds.), Game Production Studies. Amsterdam: Amsterdam University Press, pp. 7–26.

Stirling, E./Wood, J. (2021): "Actual History Doesn't Take Place: Digital Gaming, Accuracy and Authenticity." Games Studies, 21(1), n. p..

Street, Z. (2017): "Queering Games History: Complexities, Chaos, and Community." In: B. Ruberg/A. Shaw (eds.), Queer Game Studies. Minneapolis: University of Minnesota Press, pp. 35–42.

Tang, D. T. S. (2011): Conditional Spaces: Hong Kong Lesbian Desires and Everyday Life. Hong Kong: Hong Kong University Press.

Tang, D.T.S. (2012): "An Unruly Death: Queer Media in Hong Kong." GLQ: A Journal of Lesbian and Gay Studies 18(4), pp. 597-614.

Tsang, S. (2004): A Modern History Of Hong Kong. Hong Kong: Hong Kong University Press.

Von Lünen, A./Lewis, K. J./Litherland, B./Cullum, P. (eds.) (2019): Historia Ludens: The Playing Historian. London: Routledge.

Unger, A. (2012): "Modding as Part of Game Culture." In: Fromme, J., & Unger, A., Computer Games and New Media Cultures: A Handbook of Digital Games Studies. Dordrecht: Springer, pp. 509–523.

Welch, T. (2018). "The Affectively Necessary Labour of Queer Mods." Game Studies 18(3), n. p.

Wesley-Smith, P. (1982): "The MacLennan Inquiry." Hong Kong Law Journal 12(1), pp. 1–3.

Wong, A. K. (2018): "Including China? Postcolonial Hong Kong, Sinophone Studies, and the Gendered Geopolitics of China-centrism." Interventions 20(8), pp. 1101–1120.

Yeo, T. E. D. (2018): "Media and Population in Hong Kong." In: Y. Huang/Y. Song (eds.), The Evolving Landscape of Media and Communication in Hong Kong. Hong Kong: City University of Hong Kong Press, pp. 251–266.

Yue, A./Leung, H. H. S. (2017): "Notes Towards the Queer Asian City: Singapore and Hong Kong." Urban Studies 54(3), pp. 747–764.

Yuen, K. K. (2003): The Writing of Competing Histories of Hong Kong, With Special Reference to the Perspectives from Britain, Mainland China, and Hong Kong (Doctoral Dissertation). The University of Waikato.

Calculated Actions
How Game Code Makes Arguments About the Past

James Baillie

Abstract

The analysis of history in games often starts with visual and narrative presentations which are some of the clearest representations of historical topics in ludic formats. These are, however, not the only ways through which players of games access ideas about the historical past. Digital games also present implicit arguments about the sorts of pasts we imagine through their code and mechanics, which can both responsively reshape the visual and narrative elements of a game and can provide a representation of processes, dynamics and decision-making in historical and pseudo-historical settings. In settings inspired by history, whether explicitly 'historical' or not, these ideas enter a complex space where historical research, popular historiography, modern ideas and ideals, ludic imperatives and the pseudohistorical fantastic can all collide.

This paper analyses numerous examples of how access to imagined pasts in game worlds is mediated by computer code, and how code models particular historical processes. The implicit arguments that developers make about their game-worlds can provide key elements for how players intuitively feel historical and pseudo-historical game worlds should work – including but not limited to perceptions of how trade and exchange function, the role of violence in pre-modern societies, or how pre-modern people thought about their world. This feel and the space for interactions in a particular setting embed historiography at the heart of a game's processes. Players' ability to access curated selections of historical ideas through games should, therefore, be understood through the reshaping of game spaces around the player that code allows.

Keywords

game studies, medievalism, programming, code, historiography, game design

1. Introduction

Games, when utilising the past, represent and draw upon a range of elements from historical periods. The visual aesthetics of the past are immediately visible in many games: even outside specifically historical games, weapons, building styles, and dress styles often evoke elements of the past. That past is not, however, just something one can understand as an exhibited collection of such artefacts. History encompasses both the processes of historical development, and the narratives of memorialising those processes: it is a study not just of what but of why, of how people chose to act in historical contexts and the aggregate impacts and effects of those choices. Games that utilise historical or pseudo-historical ideas often cannot help but embed ideas about how and why some historical dynamics functioned, even if their objective is fundamentally about using the past as an aesthetic rather than a deeper inspiration for a historical simulation. These ideas and arguments may be made as intentional inclusions in a work, but are often the result of assumptions or the by-products of systems made for other purposes.

A lot of arguments about the past across media in general are implicit rather than explicit, in line with the broad concept of 'showing rather than telling'. The lack of an explicit statement of an argument does not, however, mean that the argument is less present or less impactful. Indeed, in some ways, the use of an unspoken assumption within or instead of an explicit argument embeds an idea much more forcefully in a work than a more argumentative structure would achieve. Such an idea becomes an expected underpinning that, unlike an explicitly presented argument, does not invite questioning or debate, but presents itself as a feature natural to the context and thus outside the scope of argument over how the world might work.

This paper seeks to look at how these implicit arguments underpin the historiographical positions embedded in digital games. It does not seek to suggest that these historiographies are necessarily intentional or that many game developers seek to directly present particular theories of the past through their games. Developers – here taken broadly and including designers as well as coders in the pure sense – are considerably more likely to take decisions for ludic or narrative effect. The full scope of how particular decisions are made is well beyond the scope of this paper, and would require a far broader overview of decision-making within development teams than can be provided here. This paper instead focuses directly on the code as a source of meaning (Marino 2020: 29–33). It examines how encoding functions in practice to communicate ideas, intentional or otherwise, and outlines how elements of the encoding process can be used in the analysis of historical arguments made by games. Design and programming decisions taken for gameplay reasons can still produce and transmit particular ideas about the way a game world works, even where the transmission of ideas is not the reason for such a decision. The resulting collection of ideas can present a wider sense of how a game's world works. Especially when this applies to societies and problem-spaces that the player is inhabiting, these can be understood as the implied historiographical positions that the game takes.

These considerations should not be confined to games that explicitly sell themselves on the basis of presenting historical topics. Whilst some writers take historical gaming to be a spectrum between different ways of representing the past through either modelled process or presentation of specific facts (Uricchio 2005: 328–331), cultural memory and the public understanding of history are not confined to the intentionally historical at either end of that spectrum. Explicitly historically inspired games often include fantastical elements, from the Amazons of Themiscyra appearing in *Rome Total War* (Creative Assembly 2004) to the power cults and Atlantean technologies of *Assassin's Creed: Odyssey* (Ubisoft 2018). Furthermore, fantastical works often explicitly draw upon and interact with concepts from historical imaginaries and beliefs. These may not seek to represent the past directly, but do represent characters, places, ideas, monsters, and deities that root those works in a framework rightly understood as connected to the historical past. Most game worlds are, then, a collection of elements which likely include some taken from historical imaginaries and some taken from elsewhere, as a curated collection.

The fictitious nature of many game settings does not free them from the world of historiography. Whilst a fantasy setting can invoke supernatural or magical logic, the suspension of disbelief often requires that fantasy operates in a way that will accord with player expectations. Those expectations will be shaped in no small part by the player's existing understandings of historical dynamics, and players may further shape, or reject, stories told by game-worlds in order to align the game's stories with their understanding of worlds and how they can work.

The process through which these historiographies and dynamics are constructed will be the core focus of this paper – beginning with the core building blocks used to do so, those of a game's code itself.

2. Historiographies as code

Before considering examples of how code produces specific historiographies, we should consider both the essential building blocks of code and algorithms that are used to produce games, and how those elements end up forming historiographical models.

A model, whether in historiography or in a game, is the overall mechanical construction of a world or scenario and how it works. Models for our purposes differ from simple representations in that they can accommodate multiple possibilities to which they can be compared and fitted: a representation can be the end-point of a model's process, but is not a whole model in and of itself (McCarty 2005: 29–30). For our purposes a specific written description of the Byzantine army would not constitute a model, but a text that tried to lay out an overall description of premodern armies could constitute one.

The purpose of a model for historians is generally to find a representation simplified and reduced from a specific idea of a past reality, to try and determine rules or shared features that can be applied across multiple contexts (Gavin 2014). Game models are not beholden to explaining the real world, and so may be constructed in different ways – for example, a reverse process of extending from a desired rule-set or dynamic that the player should experience and then building a more detailed model around that endpoint. They may not strictly represent any reality, so much as being designed to give the impression of a deeper reality beyond the modelled world.

Digital models can provide considerably more possibilities to test different variables at greater levels of precision and with a far larger number of interlocking processes than analogue and thought-models: this has a lot of utility for historians, though the process needs careful handling as to whether the historical information actually supports the stack of ideas on which the model is based. Frequently, we have few 'test data' samples especially for premodern historical contexts, making any attempt to consider variation in the model rather divorced from the available data. If we expand the model to encompass more historical contexts, so many possible variables differ between contexts we might wish to compare that agreeing on which factors actually matter and how they could be categorised (let alone quantified) can become an impossible maze of historiographical debating points. None of this means that historians avoid having models of expectations about the past – indeed we often imply the use of general models without realising it – but formalising such models is a process fraught with difficulty. These sorts of concerns about sourcing and historicity are less prevalent for a game developer, for whom a digital model of a nominally or partially historical context can provide a set of rules and ideas which the developer can select or reject based on their preferred outcome. The process of handling the model thus differs, but the processes the models simulate and encode may have significant similarities.

What we mean by code is somewhat complex, and games are often produced with numerous layers of interacting pieces of code. Here the term will be used very broadly, including all the instructions and structures that a computer uses to model (that is, to encode) a simulated setting, game, or environment. It should be noted that software engineers may use other definitions, for example using 'code' strictly to mean compiled code, the element of a game or system that is rendered immutable after the code compilation process, as opposed to scripts and input data that are read and utilised by the compiled code.

A game's input data structures can in some cases also be seen within this wider framework of code. Whilst unlike scripts or compiled code, data structures are generally functionally inert, they do in the broader sense encode a representation or model of particular concepts and objects, presenting a set of information about those things that is then processed by the algorithmic layers of the game. The structure and type of that data will interact with the process elements, defining what information is there to be processed: an algorithm can only take information into account if it has that information available. The categorisation of that information – for example, whether a character's skills at different crafts are treated as a list of learned specialisms, or the result of a small number of 'core' skill numbers representing a character's dexterity, imagination, and suchlike - can have significant knock-on effects into the rest of the model. In our example, if we use a core-numbers model, the implication may be that labour is largely non-specialised and a sufficiently dextrous character can shift from weaving to ceramics without too much difficulty, whereas a list of learned specialisms would imply a far less flexible ability to move between tasks.

The scope of what code can do is very wide, but some broad categories of its common impacts upon game models can be summarised here so that we can refer back to them when considering later examples. In brief, the categories most important to us here are code as model, code as model weighting and inputs, and code as a randomisation function.

First and foremost, code determines the basic model that the game uses for simulation elements, and includes the algorithmic functions used to make decisions and calculations. Which factors are taken into account when determining what will happen next, and how they interact to create an outcome, are at the core of determining how player choices interact with the rest of a game.

Code can, secondly, influence the weight of different factors in a model, which elements are treated as relatively more or less important. In a sense, such weighting is in itself simply another part of the model, but it deserves particular mention here because the logic of which factors are included and how they are compared can be somewhat separated from the way in which argumentative and narrative outcomes can be subsequently influenced by changing the relative scale of different factors. This may be done at the level of the game's data, or by changing the weight of a factor within a particular calculation.

The third key impact of code is that it provides randomisation functions. Some games, such as point-and-click adventures, puzzles, or choose your own adventure narratives, can work without any chance-based element, but a large percentage of games that provide a medieval inspired setting are based in genres that include calculations of weighted chance. These, as will be seen later in this paper, can influence the presentation of the setting in numerous ways, with increased or decreased randomisation changing the narrative feel of a game significantly.

Having these three elements available is crucial to the expression of historical argument through code. If we broadly consider that much historical argument attempts to answer the question 'why did X happen', then we see the modelling impact of these parts. The first defines which factors and prerequisites we take into account in why X happened, the second helps assign their relative importance, and the third is the impact of contingency. Whilst not all of historiography can be quite boiled down to a list of factors, weights, and contingent chance, these essential elements do provide considerable power for expressing historical processes.

The study of the encoding of a game-world is very closely related to, but not necessarily entirely the same as, the study of game rules in ludic contexts. A given piece of a game's code and its simulation of a game world may affect the ludic mechanics through which the player interacts with the setting, but will not necessarily do so: these elements are expressed through code, but so too may be a range of background elements of a game world. Code is regularly used to manipulate these, especially in more simulationist settings, and can therefore contribute to the game through emotional and narrative impacts as well as the stricter sense of *ludus*, or game as mechanical challenge.

On this last point, code as story has been a point of some debate, at least insofar as it begs the question of whether game code can construct what might properly be termed narrative (Ryan 2023). This is largely a definitional question: if one requires a narrative to have a set structure produced by a singular guiding hand, fewer games will fulfil the requirement than if one only requires that the player's experience involves understanding their presence in the game-world in narrative, story-driven terms. This paper will consider narrative more in the latter sense: whilst the focus of this paper is on how developers construct code that produces historiographical argument as an output, developers usually write such code to produce effects in gameplay and stories for players. They may do so, too, with the understanding that players are likely to provide further narrative in their own interactions with the game-world, building their own understandings of their space by creating narratives around themselves.

These understandings of code and rules as a rhetorical or argumentative format reflect existing usages in software and game studies. Ian Bogost, with a focus largely towards games with greater intentionality in their procedural messaging, suggests that games provide a space for procedural rules to become a form of rhetorical argument (Bogost 2007: 28–29). There is however a potential gap between intentional rhetoric and the arguments that appear in a final work: these expressions in code may be more or less lacking in visibility for the reader, and may communicate more or less complex ideas at their endpoints compared to the underlying intricacy of the code (Wardrip-Fruin 2009: 4–15).

Fundamentally a historiographical framework and any sort of code model based thereupon tends to assume a certain sort of story about human societies and their development, or at least about certain aspects of that process. Even historiographies and models that assume a high degree of contingency can produce certain sorts of narrative impressions: creating the impression of a high level of contingency in stories can in itself be a narrative effect, increasing the player's sense of anarchy, chance, and threat. In the following sections we can see the use of these core elements through a range of examples. These are approached through a mixture of close reading methods, either direct (reading the code directly) or indirect (using the game's endpoint and play to discuss the rules expressed in its underlying code): in most cases the latter is necessary when discussing major commercially published games, as these rarely make their code fully open-source (Bogost 2007: 62–64).

3. Calculated assumptions

Many examples of ludic historiographies we can look at involve points where code produces the underlying rules of a society and its interactions. As noted earlier, these become assumed norms about their game-worlds, embedded beneath the layer of elements that are questioned and foregrounded by the story.

Perhaps the simplest initial example we can use is one involving historical trade. In the real-time strategy games *Age of Empires* (Ensemble Studios 1997), *Age of Empires II* (Ensemble Studios 1999), and *Age of Mythology* (Ensemble Studios 2002), conducting trade involves a mobile trade unit moving between specific buildings owned by the player or an ally, which provides a way of obtaining gold. This is especially important in the later parts of a game, when finite gold mines placed on the map will be running out. Crucially, trade is made more lucrative according to distance (that is, any given percentage increase of distance gives a greater percentage increase in the gold resource gained).

This mechanic encodes a range of assumptions about historical trade. It presents it as being primarily about gold for profit, which can then be utilised by the player/polity. It also presents it as being controlled effectively by the state: the state (in the form of the human or computer player) controls the number of trade carts, where markets exist, and when trade should start and stop with a very high degree of precision. Notably there are *never* merchants who come from outside the immediate political situation.

We can see the results of these assumptions on the presentation of the situation better if we think about how else the system could work. For example, in historical texts we see writers such as the Byzantine strategist Kekaumenos giving a very different portrayal of how trade and warfare interact (Kekaumenos 2013). In his *Consilia et Narrationes* he writes about methods for interviewing incomers from elsewhere to gain information, and discusses the benefits and dangers of gift-giving and trade with one's immediate rivals, for example explicitly advising against banning likely enemies from setting up markets but noting how to protect against these being strategic cover for an attack on a fortress.

In the *Age of Empires* system, there is both no need to actively prevent, and little incentive to attempt, trade with an enemy: enemies will attack one another's buildings, merchants and civilians upon sight with the objective being to wipe all enemy controlled assets off the map. Capturing settlements and points rather than destroying them is utilised in some games in the series, but minimally: in *Age of Mythology*, player settlement centres can only be founded at certain 'settlement' points, and in *Age of Empires III* (Ensemble Studios 2005), 'Native' settlements outside the player's control may be brought into play by constructing a single 'trade post' building. Even in these cases, the points are inert and outside play unless are built on and claimed. The result is a very zero-sum system in which gains for one side require the absolute destruction of an opponent, and so in which trade is not a way to reach out to the world, but something that needs to be protected from it.

We could, however, envisage a game similar to *Age of Empires II* that took Kekaumenos' positions, or a different historiography of trade, more seriously. Trade carts could act as 'neutral' units that were not attacked on sight, and provided line-of-sight for their owning player but resources for both players on a route, leading the player who did not build the cart to face a trade-off between profiting from an opponent's investment on the one hand but providing the opponent with valuable information or strategic opportunity on the other, something Kekaumenos warns about with regard to gifts of food. We might also want for example to add in elements where the food resource was more important in trade where players were building advanced city structures, to model the importance of grain supply to larger urban centres, which was also a concern for some medieval writers. A different historiography of trade could therefore provide a different set of possible mechanics, encoding different ideas about the past but also changing the possibilities for gameplay.

Another historiography can be found in combat mechanics. Battles in games are frequently exceptionally deadly affairs, ended when only one side has any members still alive. This is especially the case in the RPG genre, where in games like *Baldur's Gate III* (Larian Studios 2023) or *The Witcher III: Wild Hunt* (CD Projekt Red 2015), the player can sometimes flee a battle but their opponents will almost never do so unless in a rare scripted interaction. RTS games also frequently have no mechanic for fleeing troops: troops in *Age of Empires* and *Age of Empires II* will always fight to the death. The *Total War* series, meanwhile, does have routing troops, but still has extremely lopsided casualty figures. In reality, a quarter of troops lost in a battle might be more than enough to end a force's fighting effectiveness, and losses of half would be a catastrophe, as for example Rogers (1998: 235) notes when examining western European medieval contexts.

We can see all of our three elements of code-historiographies throughout the minutiae of combat mechanics: for example, 'hit-point' systems that change the player from entirely capable at any hit point number greater than zero to entirely out of action upon hitting the zero line are the standard in both role-playing and strategy games (Peterson 2012: 331, 337–338). The mechanic was inherited from their tabletop counterparts, and has the gameplay benefit that the player avoids negative spirals where wounds or exhaustion weaken their characters and make combat situations progressively more frustrating as defeat becomes more likely. Weighting can then make further differences: by tweaking numbers of hit points or attack strength as model parameters, one can change combat from a rapid, deadly affair in which not getting hit and striking first are the primary objectives, through to combat being a slow, exhausting process in which wearing down your opponent and maintaining stamina for the duration are more important. How much of the result is random can likewise be factored in.

The presence or lack of certain things in a game's model has narrative impacts, because certain things can only feed on into the story if they are included in the mechanics. Characters wounded or disabled by battle are rarities except those presented with an essentially cosmetic scarring or amputation that has no gameplay effect, such as the characters of Beast in *Divinity: Original Sin II* (Larian Studios 2017) or Neve in *Dragon Age: The Veilguard* (BioWare 2024), whose missing eye and lower leg respectively contribute to their character aesthetics but are otherwise fairly irrelevant to and rarely commented on in gameplay. Disabilities among non-player characters can have them as figures whose infirmities render them entirely incapable, but practically disabled characters are unusual.

All of these issues change a game's historiography of combat, of what wins battles and wars. Games tend to prioritise a view with a relatively low role for contingency – there will be some randomisation, but the chance of an individual character being killed unexpectedly by a far weaker opponent who happened to hit them in the wrong spot or by stumbling on a stronger opponent is usually kept quite low unless as a mechanism for warning players of certain areas or opponents designed for a later point. Players expect combat to test their skill (whether strategic or action-based), rather than for it to be something that they might want to avoid. There are exceptions to this, as anyone who has bumped into a radscorpion pack too early in a playthrough of *Fallout* (Interplay Productions 1997) can attest: but here the pseudo-historiography supports the narrative of an especially harsh and unforgiving environment.

There are extents to which adaptations to warfare are ludically necessary: games are not, and cannot be required to be, 'accurate' in this regard or any other. As noted above, the first aim must be to understand *why* certain processes are represented in certain ways, and why certain aspects of those processes are left out. A large aspect of this is the reduction of narrative and gameplay friction. The finality of killing opponents is simpler to represent on-screen than injury and retreat, especially in large virtual spaces where an enemy might have to run a long way to leave the play area. It also avoids the player being confronted with the unsettling experience of wounded or traumatised enemies. The result of this is that games tend to present a medieval or pseudo-medieval world that is far, far more deadly than the results of actual battles, and in which the results of battles are often a sharp binary between total victory and total defeat.

This heroic historiography of combat, where skill at arms or generalship is central to victory and victory is a simple matter of cleanly deceased enemies, understandably owes more to the history of epic literature than the history of warfare. It is a narrative tool, focused upon taking combat and using it to tell a story with a protagonist overcoming a particular challenge. This modern popular historiography of war is not just represented in games - it is often central to how people think about the medieval period. Many national historiographies put great weight on battles as historical turning points: Las Navas de Tolosa, the Battle of the Neva, Covadonga, Agincourt, Hastings, Basiani, Didgori, Tours, Grunwald, Ain Jalut, and Bannockburn are just a handful of specific medieval examples. These battles are treated as politically and experientially formative for their respective polities: a player experiencing these or evocatively similar moments in games places themselves to an extent within those formative frameworks and within certain ideological and historiographical views of how polities are imagined to form and grow.

Gamers are unlikely to cite games, especially fantastical ones, as specific evidence for historical claims about how battles were won: it would be wrong to assert too simplistic a linkage from a game to assumptions about reality. Nonetheless, games are still likely to have an impact on people's initial expectations and what will feel plausible or authentic to them: they build up cultural memory of struggle, misplaced or otherwise. The presence of the Latin phrase *Deus Vult* in modern far-right circles can in part be linked to its popularisation through the *Crusader Kings* series (Bishop 2019). Whether presented as valiantly heroic or gritty and bloody, the feel of battle matters, and is heavily delivered by the code as much as the visual layer of a game.

We can, then, end this section by considering how else a feel of combat might be used. A wide array of balances and systems are used across different games, but these are often quite genre-restricted. A stronger set of mechanics for enemy retreat and surrender might be an option in both role-playing and strategy games that could provide rather than restricting narrative possibility, giving players meaningful choices. Modes of play that more effectively centre disability in its social-model sense may be difficult to produce in certain genres while retaining players' desire for agency. Such play might, however, permit portrayals of historical violence that allow for its existence but take a more critical view to its glamour, whether in a sanitised heroic form or in darker tones that nonetheless still centre the assumed necessity of extreme violence. Allowing play that engages with social structures and systems for the avoidance of violence could open more avenues for storytelling through role-play that would make little sense when paired with more conventional combat mechanics.

4. Calculated aesthetics

Our examples in the previous section were primarily ludic, and involved mechanics that the player might interact with directly. As noted in the introduction, however, the study of code as historical argument need not confine itself to purely interactable ludic mechanics. In many games that involve simulating a historical or fantastical world, aesthetic elements are crucial to the play experience. Whilst in a sense, all game aesthetics involve code to render to the player, here we will mainly consider situations where the aesthetics specifically adapt or are manipulated by code in response to gameplay situations or player behaviour.

Aesthetics should also not be treated, in these cases, as a sealed separate box to the gameplay: they represent a significant influence on how certain choices are framed, and players often attempt to play games in ways that are narratively satisfying rather than simply trying to 'beat' a game or reach an end screen efficiently. This may include playing to produce particular assemblages or situations that feel historically or aesthetically correct, a process Chapman dubs "configurative resonance" (Chapman 2016: 43–46). How a player feels about the game world is likely to influence how they play the game, even if the game is making substitutions that are mechanically irrelevant or providing feedback that has little direct impact on the gameplay.

An example here is using code to provide visual variation. In games where the characters and settings are not all pre-defined, variation in the look of units can involve code systems to swap collections of parts or 'skins' for each model. In the *Total War* series, for example, early games in the series have all members of a unit of troops looking identical: later titles, from *Medieval II: Total War* (Creative Assembly 2006) onwards, provide visual variation in the exact look of each 3D model by using code to randomise which set of texture elements are applied. This change has no gameplay effect at all: the characters in the unit are still identical in ludomechanical terms. Games usually apply significant weightings to variation in looks, as with the complex 'DNA' systems through which *Crusader Kings III* (Paradox Development Studio 2020) characters are generated to look like some variable mix of their parents' physical traits.

This sort of aesthetic code is especially important in that debates over visible diversity in medieval and pseudo-medieval games are a major point of contention in popular culture and reception of the medieval period. Medieval and medieval-fantasy games have been criticised on numerous occasions for failing to present a diverse array of racial and ethnic groups, or conversely for portraying ethnic groups that certain players do not see as fitting into their aesthetic imagination of medieval Europe. Who is permitted to see themselves in a pseudo-medieval world, and who can fit themselves into modern perceptions of who owns certain elements of culture or heritage, may be directly affected and influenced by who people can see presented in related art forms such as games.

Another example of a purely aesthetic element that may be manipulated by code would be 'idle' elements of a game. In many RPGs, characters are portrayed doing a range of activities when the player is not immediately interacting with them, often referred to as an *idle animation*: they will usually stop these activities if the player initiates an interaction, such as conversation or combat. The location of a non-playable character (NPC) if manipulated by code, might have a marginal ludic effect by changing the precise location of a combat or other such encounter, but in large part these are aesthetic presentations with minimal gameplay impact. They are nonetheless important to the game's storytelling: a character shown cooking, versus one shown drinking wine, versus one shown sharpening a blade, all tell the player about each character and the wider game world. The balance and choice of these activities may be manipulated through code, influencing that presentation.

These examples show code presenting different understandings of the past, without directly impacting upon the ludic elements of the game. This may indirectly interact with the game's mechanics if it changes the ways that a player approaches the game by providing them with certain sorts of information, or indeed by encouraging certain sorts of player-defined play and goals.

Narratives can equally be manipulated aesthetically. Whilst some choices in games do have mechanical impacts on the flow of the game, there are often other choices or narrative elements that have less ludic impact and are predominantly aesthetic in nature. Obtaining certain aesthetic choices may become a game objective in itself for some players, adding a layer of complexity as players redefine the problem-space of a simulation game for themselves.

An extreme example would be in the game *The Exile Princes* (Pangolin Games 2024), in the 'tunnels' section of the game: in this case, the player makes a number of 'decisions' about their direction of travel when moving through an underground environment beneath one of the game's cities. The game has no graphics or other frame of reference for the player to test their decisions against, nor are they given the option to double back to check another option. This is because the decisions are mechanically meaningless: they are randomly selected and simply provide a rough aesthetic of exploration between mechanically significant random events such as discoveries or combat encounters. The manipulation of these or the addition of different factors via code can nonetheless provide feedback to the player on the kind of world that they are inhabiting: the balance of particular decision frames is influenced by which enemy will be faced if the player reaches the end of the tunnels. Similar systems could add to a game's historical position in non-ludic ways. What a player gets shown at what point can help define the player's perception of a world, and thereby perhaps their expectations of how to interact with it.

Sometimes, this can come by overriding a game's usual processes. Take for example the game *Hades* (Supergiant Games 2020), set in a reimagining of the classical Greek underworld: the broad objective of the game is to break out from Tartarus and reach the surface with the help of various Olympian deities. In general, the order in which one finds deities' 'boons' that boost the player character has strongly randomised elements, which along with randomised room order and death forming a regular game mechanic place *Hades* in the *roguelite* genre. There are exceptions, however, to support the narrative and aesthetics being revealed at the appropriate times. For example, Athena is predetermined as the first deity the player meets with a new character. This in part fits with Athena's mythic role as a patron of heroes, but may also be connected to Athena's in-game presentation as a black woman: this deviates from typical 19th and 20th century western presentations of Athena, and helps set the tone for the player that this is an interpretation of myth that is willing to look outside conventional portrayals. Encoding certain orders of events, even if not fundamental to the ludic core loop, can produce narrative emphasis and therefore display the game's themes or positions on given issues.

This mix of calculation and randomisation forming a historiography can, then, apply to manipulating the sensory and visual world of a game as well as its strictly mechanical level. Elements that do not contribute to the ludic, interactive sense of winning a game can still affect how a player approaches a game. What people expect past worlds to look like, and the game's interaction with those expectations, informs the game's sense of place and therefore its implicit historiography. The specifically dynamic construction of a game's setting and aesthetics through code can therefore also be treated as a form of encoded historiography.

5. Calculated actions

The third case study area to consider will be how game code models decision-making processes specifically. This is an area that is especially amenable to reflecting historiographical positions. The factors in why decisions were made, and who made those decisions, are fundamental to the understanding of the human past, though also among the more unknowable elements of that past.

A first example might be, in role-playing games, the scripting of non-combatant NPCs with regard to combat. In some cases, for example *Divinity: Original Sin 2*, non-combatants will almost always be scripted to move as far away from the combat as possible, sometimes with cowering or fleeing animations to visually display that status to the player. In other games, for example *Skyrim* (Bethesda Game Studios 2011), even generally passive and clearly noncombatant NPCs may counterattack in any case of a threat to them or their fellow innocents.

The calculations these NPCs make may not be made for historiographical reasons, and indeed almost certainly are not: if a developer for gameplay reasons prefers to have NPCs out of the way of any violence, for example making it more likely that the player will be able to interact with them later or simply to convey the emotional sense of panic at the situation, then it makes sense to have them run. Conversely, the player being swamped by NPCs responding to a threat could be a disincentive when it comes to player action: attacking guards and regular NPCs in the *Elder Scrolls* games for example is meant to be sufficiently frustrating for the player that it discourages random violence on the part of the player.

In the case of when noncombatants choose to participate in violence, the calculated actions nonetheless say something about the world the game happens in. Whatever the ludic drivers for action, the fact that a resulting historiography is unintentional does not make it non-existent. Whether random NPCs are prepared for and capable of violence tells the player something about expected norms for violence in a game-world. This is an implied historiography: the extent to which people in past societies carried and learned the use of arms varied. Expectations that a premodern world would have more widespread capacity for violence may have some roots in the realities of worlds of levied armies and hue-and-cry responses, when the monopoly on violence was less confined to a small, professional segment of the population than is usually the case today. Such expectations also, however, reflect post-medieval ideas of a 'dark age' that can be compared to its more 'civilised' counterparts in other periods. What players will consider reasonable likewise depends in part on their wider perceptions of a game world, which are grounded in such historical and cultural ideas.

How games calculate actions can also be a matter of group or polity behaviour – for example, numerous games will include a set of diplomatic mechanics that influence making war and peace between realms. Here, too, historiography and code clearly intersect. To what extent, a historian might ask, did Byzantine rulers' dalliances, amiability, or cruelty impact upon the overall policy and actions of the Empire (Angold 1984: 263–274), or English monarchs' attitudes and angers on the belligerence of their armies (Spencer 2017)?

A game like *Crusader Kings III* attempts its own mechanical answer to the question, in which for example a more skilful general or a more bellicose or ambitious leader is more likely to take aggressive actions, whereas a more contented one is less likely to do so. Characters have a trait system where 3-4 personality features are noted – being 'craven', or 'ambitious', or 'generous', among other examples – and these can then inform strategic decision-making. These factors are balanced against 'core' pressures: in *Crusader Kings III*, having an heir and state survival might be treated in this regard, so rulers will act against their personality to avoid losing land or their dynasty and there is no way to prevent them caring about those core factors.

This approach tends to a personality-driven historiography in which the personal attributes of a ruler are key to wider historical developments. Other positions – and mechanics – are certainly possible. Games may take the attitudes of people other than the monarch into account: *The Exile Princes* factors in every noble who owns a city equally into a faction's strategy, which can lead to effects being balanced out or multiplying depending on the balance of leadership. Conversely, a game like *Civilisation II* (MicroProse 1996) might not implement individual rulers at all, in favour of a grand strategy narrative in which the faction as a whole has a single 'personality' from a strategic perspective without the temperaments of individual rulers affecting a faction's strategic outlook.

As with our other choices, the code layer affects what can be done here. The *Civilisation* series (not least because of its large time jumps) does not model a faction's political elites as individuals, so they cannot be taken into account when considering the faction's strategy. There tends to be close alignment between the data-as-code and the model here: information can only be mechanically in the model if it is included in the data, and the data will usually be structured with particular processes in mind.

These are choices that interface quite directly with certain historical debates. Few historians would espouse the idea that longue-durée civilisations had strong inherent tendencies to aggression, peaceability, or other such traits, though this viewpoint was influential historically: it was crucial in the development of, for example, modern western Islamophobia, and still permeates elements of public discourse especially on the far right (Wollenberg 2014). The balance between a polity being modelled as a network of elites, as reliant mostly on its leader, or as fundamentally a singular entity tending to a certain corporate or cultural 'personality', is certainly something on which actual historiography is as varied as ludic approaches. Games tend to seek legible, time-efficient ways to explain their mechanics to players, which may tend to lead to approaches where polities act in more similar ways to one another or have their outlook anchored in a single representative individual.

As a final example, we can look at the interaction of calculated action with something often treated as less mutable than an immediate problem space: identity. Identities are often quite rigidly locked down in games: it is rare that characters are presented with flexible identities, and indeed medievalist fantasy has a long history of speciating certain groups. Choice is less often taken into account in games when presenting non-player identity, and indeed player identity after the creation of a character, although the presentation and performance of certain identities undoubtedly affect people's reactions to one another throughout historical societies. Sometimes identity can be a scripted narrative question for a character, notably Daelan Red Tiger in *Neverwinter Nights* (BioWare 2002) or Taash in *Dragon Age: The Veilguard* (BioWare 2024). These examples essentially make the different possible identities an inherent conflict for their characters, though, creating a precise choice of end-states: whilst the player may get to influence which identity dominates, there is no way to approach these characters' identities in a way that takes full account of the fluid and situational character that identity often has for individuals past and present.

Such crystallised, simplified identities are easier to manage in data and more convenient to write as clear narrative pathways, but potentially provide a weaker sense of historical identity formation and performance. How such identities are produced in data matters enormously for what sort of historiographies a game can have in it, especially in games that have a strong simulation component. There are often significant historiographical debates over which factors – language, dress, social networks, prayer, or myriad others – were most taken to represent certain identities historically: in medievalist games, however, the norm is that everyone has a single, accepted ethnocultural identity that is obvious to all around them. This then feeds into actions and interactions: in some games for example, characters will interact more readily with those who are culturally familiar to them, as in for example *Morrowind* (Bethesda Game Studios 2002). Such mechanics would be complicated, but potentially made more interesting, via a more complex code-level expression of the constituent parts of cultural identity.

By understanding the process of calculated actions in games we can, then, begin to understand how to recalculate them. This can mean challenging our ideas about how to portray past cultures in games, and avoiding revisiting familiar monoliths in their presentation – a challenging ask, given the other pressures on developers. There are opportunities, however, in finding such alternative ways to represent the past, potentially deepening the social aspect of role-playing games with less blunt representations of culture, or considering how different ways of handling the respective roles of individuals and polities might help developers express alternative ways to interact with medieval politics.

6. Conclusions

As this paper has shown, game code can make arguments about the past, and can embed them on a deep and implicit level into games both historical and ahistorical in nature. These historiographies are transmitted via the games as a medium not because their designers strictly intend to get across a particular view of history, but often because they do not have any such intention: these are the assumed norms that underlie how historical dynamics and thought processes are calculated.

We have seen this through a code-centred approach – essentially looking holistically at the dynamic elements of a simulated system and their impacts on a player. This has been subtly different from a strictly ludomechanical 'rules-based' approach or a problem space analysis (McCall 2020), not just in that it can incorporate aesthetic mechanics as well, but also in that it asks some different questions. Rather than treating the player's influence over the world and their pathway through play as the fundamental object of study, the code-centred approach used here is intended to ask questions about the game-world as a whole system from which the player gains a certain understanding and therefore a certain reflection on the pasts with which it engages.

This article has shown the potential of this approach across an array of topics for assessing the implicit historiographic character of games, including showing that the method can handle not just general historical processes as they appear in strategy games but can be applied to game aesthetics and individual or social dynamics. There are nonetheless difficulties in the application of the approach, in particular the lack of access to code-bases: as Bogost (2007: 62–64) suggests and as has been attempted here, this can be to some extent counterbalanced by effective reading of processes from a player perspective, but that in turn leads to questions about how well the mechanics as they appear to the player reflect the underlying code (Wardrip-Fruin 2009: 15–16). Both for helping read the relationship between code and front-end and to allow more in-depth analysis, finding ways to access code-bases for analysis is still highly desirable. This may well mean working with smaller developers and encouraging engagement outside the AAA game markets where code access is likely to be most jealously guarded.

There are numerous potential future directions in which this sort of approach and research could be taken. Taking this approach more systematically with specific games could help to understand the interplay between how developers perceive player expectations of the past and the role of a game's coders in building a 'feel' for imagined past environments that can in turn feed into the wider cultural and political space of game related discourses. Taking fully creative approaches based on the ideas in this paper should also be encouraged both as ludonarrative art and as research practice, exploring the formalisation and creative use of our cultural memory of historical processes to better understand their replication and construction.

One question that this paper and this approach fundamentally cannot answer, on the other hand, is the total impact of this historiographical embedding. Some of these elements might be used as signifiers of overall 'historical accuracy' or precursors to feelings of 'authenticity', both terms in common usage for assessing historical games (Kapell/Elliott 2013: 359–361) but which are often rather nebulous (Donald/Reid 2023: 63–70) or, as discussed below difficult to assess holistically. Whilst exploring those surrounding contexts and reactions in full is beyond the scope of this particular discussion, feelings of historical authenticity and players' and developers' senses of accuracy are key routes through which embedded historiographies have wider impacts. It is therefore useful as a grounding to assessments of those impacts to understand the construction and use of these elements in and of themselves, in the ways that this paper has attempted to outline.

This is especially the case when thinking about what players do find authentic in pasts and pseudo-pasts shown to them in games. Studies of the wider impact of game-driven historiographies and public perceptions of authenticity in historical-style settings have been limited. In the absence of large-scale survey work, for which game studies and history departments often lack the funding and capacity, these factors are obscured from the view of scholars. Work on these issues has largely been based upon small focus groups and classroom settings, which are important, but may not be applicable to gaming audiences more generally (O'Neill/Feenstra 2016; Stirling/Wood 2021). We should not take simplistic assumptions on this matter: there is little evidence that players take games entirely literally when thinking about, evoking, or using historical material, but we should not assume that this means they fail to take them seriously. Games and their surrounding cultural and 'affinity' spaces create areas for historical and cultural absorption, transfer, contention and debate (Hartman/Tulloch/Young 2021). We should expect players not to view games as historical sources, but to take them as parts of a conversation with the past and its impact on present ideas, imaginaries, and identities.

It is critical that we understand how games are constructed around these implicit arguments, despite these difficulties in understanding player perspectives directly. We cannot fully assess the totality of the public impact of a play, a film, or even a history book on people's understandings of the relationship between past and future, and the same applies to games. These analyses nonetheless allow us to understand better what messages about the past games are constructing, and do so in a granular way that shows the underlying systems to best effect. Those messages matter beyond being mere curiosities because they feed into wider discourses of cultural memory and identity in a world where those things are part of very present political polarisation and violence.

Analyses of this kind also allow the historian and developer to step into one another's shoes. Understanding how code makes argument is an act of translation between humanistic understanding and creative technical work, and – *traduttore, traditore* - this might sometimes be uncomfortable both for the historian seeing the formalisation of rules of thumb they would not always admit to using and for the developer seeing the histories and narratives they did not know they were writing. Nonetheless, it creates opportunities that are worth considering – introducing a wider array of historical ideas into games and working on how people experience historical processes through game-worlds could be a source of opportunity for developers and historians alike. Translation between art and history is a betrayal of each to the other precisely in that it is an act of creating something new. It is, if pursued, a process that could provide creative, constructive approaches to helping people engage with the calculated actions of people in the past and with the processes and systems from history that we choose, individually and collectively, to remember.

References

Angold, M. (1984): The Byzantine Empire 1025-1204. New York: Longman.
Bethesda Game Studios (2002): The Elder Scrolls III: Morrowind [Video Game]. Bethesda Softworks.
Bethesda Game Studios (2011): The Elder Scrolls V: Skyrim [Video Game]. Bethesda Softworks.
Bishop, A. (2019): "#DeusVult." In: A. Albin/M. C. Erler/T. O'Donnell/N. L. Paul/N. Rowe (eds.), Whose Middle Ages? New York: Fordham University Press.
BioWare (2024): Dragon Age: The Veilguard [Video Game]. Electronic Arts.
BioWare (2002): Neverwinter Nights [Video Game]. Infogrames.
Bogost, I. (2007): Persuasive Games: The Expressive Power of Videogames. Cambridge: MIT Press.
CD Projekt Red (2015): The Witcher III: Wild Hunt [Video Game]. CD Projekt.
Chapman, A. (2016): Digital Games as History: How Videogames Represent the Past and Offer Access to Historical Practice. New York/London: Routledge.
Creative Assembly (2006): Medieval II: Total War [Video Game]. Sega.
Creative Assembly (2004): Rome: Total War [Video Game]. Activision.

Donald, I./Reid, A. (2023): "Account, Accuracy, and Authenticity: A Framework for Analysing Historical Narrative in Games." In: R. Seiwald/E. Vollans (eds.), (Not) In the Game: History, Paratexts, and Games. Berlin/Boston: De Gruyter.
Ensemble Studios (1997): Age of Empires [Video Game]. Microsoft.
Ensemble Studios (1999): Age of Empires II [Video Game]. Microsoft.
Ensemble Studios (2005): Age of Empires III [Video Game]. Microsoft Game Studios.
Ensemble Studios (2002): Age of Mythology [Video Game]. Microsoft.
Gavin, M. (2014): "Agent-Based Modeling and Historical Simulation." Digital Humanities Quarterly 8(4), n. p.
Hartman, A./Tulloch, R./Young, H. (2021): "Video Games as Public History: Archives, Empathy and Affinity." Game Studies 21(4), n. p..
Interplay Productions (1997): Fallout: A Post Nuclear Role-playing Game [Video Game]. Interplay Productions.
Kapell, M./Elliott, A. (2013): Playing with the Past: Digital Games and the Simulation of History. London/New York: Bloomsbury.
Kekaumenos trans. Roueche, C. (2013): Narratio et Consiliones. SAWS edition.
Larian Studios (2023): Baldur's Gate III [Video Game]. Larian Studios.
Larian Studios (2017): Divinity: Original Sin II [Video Game]. Larian Studios.
Marino, M. (2020): Critical Code Studies. Cambridge, MA: MIT Press.
McCall, J. (2020): "The Historical Problem Space Framework: Games as a Historical Medium." Game Studies 20(3), n. p.
O'Neill K./Feenstra, B. (2016): "'Honestly, I Would Stick with the Books: Young Adults' Ideas About a Videogame as a Source of Historical Knowledge." Game Studies 16(2), n. p.
McCarty, W. (2005): Humanities Computing. New York: Palgrave Macmillan.
MicroProse (1996): Civilisation II [Video Game]. MicroProse.
Pangolin Games (2024): The Exile Princes [Video Game]. Pangolin Games.
Paradox Development Studio (2020): Crusader Kings III [Video Game]. Paradox Interactive.
Peterson, J. (2012): Playing at the World: A History of Simulating Wars, People and Fantastic Adventures, from Chess to Role-Playing Games. San Diego: Unreason Press.
Rogers, C. J. (1998): "The Efficacy of the English Longbow: A Reply to Kelly DeVries," War in History 5(2), pp. 233–242.
Ryan, M-L. (2023): "Narratology for Game Studies", The Encyclopedia of Ludic Terms.
Spencer, S. J. (2017): "'Like a Raging Lion': Richard the Lionheart's Anger during the Third Crusade in Medieval and Modern Historiography," The English Historical Review 132(556), pp. 495–432.
Stirling, E./Wood, J. (2021): "'Actual History Doesn't Take Place': Digital Gaming, Accuracy and Authenticity." Game Studies 21(1), n. p.
Supergiant Games (2020): Hades [Video Game]. Supergiant Games.

Ubisoft (2018): Assassin's Creed: Odyssey [Video Game]. Ubisoft.
Uricchio, W. (2005): "Simulation, History and Games." In: J. Raessens/J. H. Goldstein (eds.), Handbook of Computer Game Studies, pp. 327–340.
Wardrip-Fruin, N. (2009): Expressive Processing. Cambridge: MIT Press.
Wollenberg, D. (2014): "Defending the West: Cultural Racism and Pan-Europeanism on the Far-Right," Postmedieval: A Journal of Medieval Cultural Studies 5, pp. 308–319.

Re-Enacting 9th Century Baghdad
Interview on the Narrative and Worldbuilding Aspects of the Past, as Rendered in *Assassin's Creed Mirage*

James Wilson, Eduardo Luersen, Raphaël Weyland, and Sarah Beaulieu

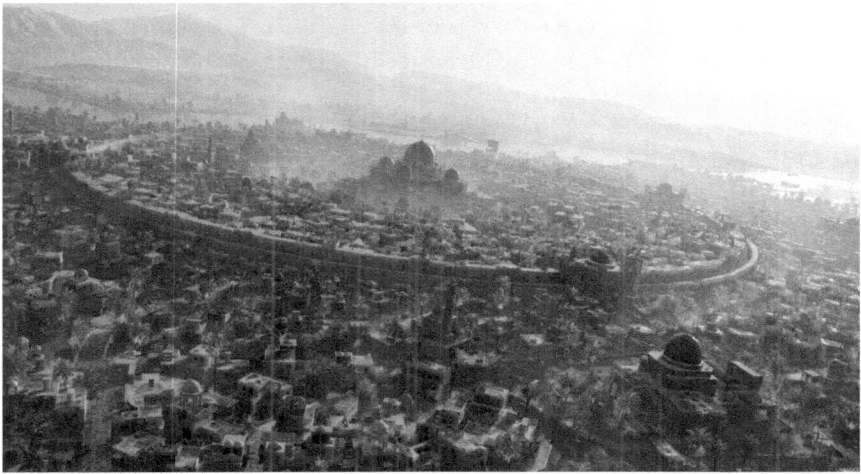

Fig.1: Screenshot of the Baghdad (9th-century) model, as designed for the promotional materials for Assassin's Creed Mirage.

Assassin's Creed is one of the world's best-selling video game series, and it is Ubisoft's most popular franchise. The use of popular historical settings is a trademark feature of the *Assassin's Creed* games, and the most recent title, *Assassin's Creed Mirage* (Ubisoft 2023), which was released in October 2023, places players in ninth-century Baghdad during the height of the Abbasid Caliphate.

We interviewed two Ubisoft employees who worked on the game and kindly accepted our invitation to discuss game production and historical aspects with us. Their insights reflect their own perspectives, which may differ from the editors'. We include them for the valuable contribution they make to the critical conversation, even when we do not fully share their views.

Raphaël Weyland is a historian who works at Ubisoft Montréal. He worked as the lead historian advising on *Assassin's Creed Mirage*. Raphaël holds a PhD from the University of Montréal, and his thesis focussed upon the city of Seleucia on the Tigris during the Seleucid and Arsacid eras.

Sarah Beaulieu was the Narrative Director for *Assassin's Creed Mirage*. Sarah is a writer for multiple media outlets, and besides writing for games such as *Assassin's Creed* and Ubisoft's forthcoming title *Beyond Good and Evil 2*, she has written scripts for virtual reality and several other formats. Sarah is also a consultant and a script reader for movies, video games and virtual reality experiences.

Eduardo: *I would like to start with a more light-hearted question. When and why did you decide to work in the gaming industry? I would appreciate it if you both could tell us a bit about how it happened, and how your previous expertise, as a historian in the case of Raphaël and as a writer in the case of Sarah, played a role in this.*

Sarah: I started working in the video game industry around 2017, after completing a Master's degree in Interactive and Transmedia Writing. My first position was in a small studio specialising in mobile games. From there, I joined Ubisoft Montpellier, where I worked as a writer. After that, I collaborated with various independent studios, working on virtual reality projects, interactive exhibitions, and similar experiences. Eventually, I returned to Ubisoft as Narrative Director at their Bordeaux studio, where I worked on *Assassin's Creed Mirage*.

As for why I chose to work in the video game industry, I had already been a screenwriter for almost a decade, primarily in linear media such as film and theatre. However, as a lifelong gamer, I had long wanted to contribute to this industry. That's why I went back to school to learn how to master storytelling specifically for video games. I also specialised in narrative design, which is about creating synergy between game mechanics and storytelling.

Today, I work as a multi-platform writer, continuing to contribute to both interactive and linear media. It was a deep desire to explore the art of storytelling in games that brought me to this industry.

Raphaël: In my case, I did a PhD in History, as you mentioned. Already before I finished my dissertation, I tried to find work as a historian, which, as you can imagine, is not always easy.

I would say that I spent the last 15 years working for whoever would pay me to be a historian. This meant working in libraries, universities, museums, different government agencies in Canada and elsewhere, or as a tour guide. I have also worked as a radio host, commentating about history. The common element in all of this has been a love for communicating history to a non-academic public.

I must say, it was not exactly what I aimed to do at the time, but since I could not find enough stability in academia, I shifted. And in a way, *Assassin's Creed* is the pinnacle of this journey – it is about using modern means to communicate about history and teach history to an even wider audience. As I understand it, my previous experience speaking to both specialist and non-specialist audiences was key to why Ubisoft chose me, rather than someone else, to work on the historical aspects, which needed to be softened and adapted.

Eduardo: My follow-up question takes us into some of these game production aspects. At which stage of the process of the development of Mirage did each of you join the production team? And, by the time you joined, which decisions had already been taken about the new Assassin's Creed title?

Sarah: In short, I was part of what we call the 'core team' – the small group of people who make the big decisions. I was there from the very beginning, and back then, *Mirage* was originally intended as a DLC (downloadable content) for an existing *Assassin's Creed* game. But eventually it was decided to develop it into a standalone title instead.

When I entered the project, it was already decided that we would do something special for it. There was a little hesitation about the city though, as we were still considering choosing between places like Jerusalem again or maybe Constantinople. For example, we knew it would take place in the 9th century because we would like to reuse a character from a previous game, who would be the main character in Mirage. Eventually, we settled on 9th-century Baghdad for various reasons, which we will probably discuss later. But yes, I was involved in making those core decisions from the start.

Raphaël: I did not participate in any of these major initial decisions; I came in later. About a year and a half before the game launched, actually – so before mid-point into the production, so not in the very early stage. By that time, many of the key gameplay decisions, such as the map size and the fact that it would be set in Baghdad, had already been made.

I worked more on smaller, granular decisions, particularly around the educational aspect of the game, like how we would try to communicate things within the game, how we would play with history, what we would change and why. At the end, every major decision is taken by the core team, but these were some of the preliminary discussions in which I was involved.

James: Sarah, how difficult was it to craft a coherent, engaging narrative for Assassin's Creed: Mirage? Did the established Assassin's Creed framework feel limiting in any way? Do you wish you could explore other facets of 9th-century Baghdad, beyond inserting famous historical luminaries into the storyline?

Sarah: As for the first question: was it difficult? Yes, of course. Working on *Assassin's Creed* is a big responsibility; it is one of the biggest video game franchises out there.

However, we had what we called an 'authenticity team', which consisted mainly of historians like Raphaël. There was also what we call the 'diversity team', which includes specialists in other subjects beyond history, such as religion, for example. We also had a 'narrative team', an 'art team', an 'audio team', and others who would gather very regularly and talk about specific topics. That would help a lot with the narrative, and that leads me to your second question: in this sense, I do not see the historical framework as limiting at all; constraints are, first and foremost, a lever to creativity.

We had a very strong relationship with the authenticity team, and that was very helpful, narrative-wise, to have new ideas on historical topics. A very concrete example was the work on the historical character Ali ibn Muhammad, the leader of the Zanj Rebellion. Initially, I knew little about him, I had only read a few lines about him in a book. Raphaël was not at the team at that point, but then one of our other historians helped me with research. That allowed me to dig into this character, but even if I would not have taken in any of the historical aspects, that would have been a lever for new narrative ideas. So, rather than seeing the historical framework as limiting, I found it inspiring.

If there is one historical aspect I would have liked to explore more in Mirage, it is the Zanj Rebellion. In every *Assassin's Creed* script, we always have a major historical conflict that we build on. And that is what ties into the ongoing and very specific lore of the Assassins and Templars. In the ongoing lore, these two sides form opposing factions of a central historical conflict, and that is something we build on. So, again, it is not limiting, it is inspiring, and we had this authenticity team to help us on this on this aspect on a daily basis.

James: Raphael, from a historian's perspective, Sarah mentioned the complexities of setting a game in a historical period like 9th-century Baghdad. Mirage not only deals with the Zanj Rebellion, as Sarah mentioned, but other controversial topics like the 'Arabic Translation Movement'. These subjects have been the subject of intense debates, especially among intellectuals in the Arabic speaking world. So, what are your thoughts on using this material to craft historical narratives and digital representations of these societies, when some individuals or groups feel very strongly about these issues?

Raphaël: I would start by saying that when we make the game, we try to create an engaging narrative. A key focus is to always provide enjoyable gameplay, because the primary objective is not to educate, but rather to entertain. The first goal is to create a game that is fun, and history in this sense is a tool to making it fun and immersive. So we try to build from settings, periods, moments that have some resonance, even if some subjects, like the American Revolution or the French Revolution are more well known than others.

That being said, even if we are not primarily trying to educate people, the more research we have available, the easier it becomes for us to produce an engaging and entertaining experience. When we have to explore a very sensitive subject, which people may expect us to engage with, then it is very useful if there is a lot of pre-existing research available as we do not have time to do all of it by ourselves. It helps when there is some resonance, as the game can draw upon things that entertains players and makes them happy. When we choose a setting, we always try to choose something that would be like that, but mostly a period of conflict is needed because, as Sarah said, the [narrative][1] conflict of the Assassins and the Templars must be presented. They have different views on humanity: Templars believe that humans are fundamentally bad and need to be controlled and the Assassins believe in freedom and that people will make the right decisions. Therefore, the game needs a historical setting in which this conflict can be situated, and from which it can draw its life.

And, in the case of Baghdad in 861 CE, there was a rebellion by enslaved people, but not only that, also the general idea of a 'Golden Age' and all that lies behind this concept. There is the Translation Movement, indeed, but why? Why is it important? Even if we do not want to educate people as a primary goal, it was very important for our team to try to attract attention to the fact that European or Western societies do not have a monopoly on knowledge, and the pursuit of knowledge. Even if we are not trying to teach people, if we can convey this idea then we are happy to do it.

In the case of the Zanj Rebellion, it was also a way to provide a setting for, not only a good versus evil conflict – but freedom versus control. As Sarah noted, even if the game does not explore every facet of the rebellion, the stakes of it are well represented and you can gain an understanding of why there was a rebellion, what was it against, and what could be the reality at that moment. The team tries to represent these pieces, these elements from history, but they are never at the forefront of the game, because they are a tool, rather than the ultimate goal.

Eduardo: *I would like to pick up on this to return to the practicalities of game production. How many people were involved in making such key decisions about the inclusion and design of historical tropes, figures, or motifs for the game?*

Raphaël: There are always multiple levels of decision-making involved but in the end, many decisions have to be taken collectively because they will affect not only one team, but many teams at the same time.

For example, if a building like a mosque, a library or, let us say, the House of Wisdom, is included, that decision will affect many teams – art, narrative, gameplay – so it requires collaboration. To use the House of Wisdom as an example: a team needs to design and set an important building into the game and it then comes to the historian for suggestions of where to locate it. We initially have no idea of what this building would look like, where it would be, or what exactly people would do in it. But we need to try and locate it somewhere in the game. From our research we knew that there was a district in Baghdad that was renowned for the production of paper, and therefore was an important place for scholars and intellectuals. Therefore, we finally suggested that this building should be included in the central narrative of the game, and were able to propose a specific location

---

1   Brackets added by the editors.

that would make sense from a historical point of view. But then the gameplay team would like to have a say in this, because at some point they would like to set some training section of the game in that very location, or would like players to ride a horse or experience any other specific gaming situation there. Then we would need to intervene and discuss, because that location would have already been taken for the House of Wisdom, and therefore the gameplay team cannot put their horse there. The art team might then also come and suggest they were envisioning a place where players could see the sunset in the game, and then you would have to say it cannot be there.

Even in a decision that you believe would make sense from a historical point of view – putting the building where it was – there are still other stakeholders involved, and because the game is a collective piece of art or creation, multiple people have to discuss and be included in these decisions. If there are too many stakeholders disagreeing, eventually one of us would have to go up to the core team, or even higher up if necessary. In the end, it may be decided that gameplay considerations dictate that a building cannot be placed in a historically accurate location. Ultimately, it is a collective process.

Sarah: Absolutely. Raphaël summarised it well. Some of the decisions would impact everybody, and the case of the House of Wisdom is a very good example. That reminds me of something very important: from very early on, every team has what we call, in the gaming industry, 'pillars'. For example, a narrative pillar of mine would be Shakespearean tragedy, and a pillar is meant to be… well, a pillar! That is, it is a vision that everybody has to follow. And if you have a question about something that you would like to incorporate in the game, you have to check if that actually fits the pillars of each of the teams – narrative, audio, game design, quest design, etc.

The House of Wisdom is a good example because very early on in the process we had to identify with the art team, and specifically Jean-Luc Sala, the art director, which important buildings we would have in the game. In the case of the House of Wisdom, it was clear that we had to include it in the game because it was very important on a historical level. I therefore had to incorporate it in my narrative, and the same goes for the art level and quest design, because at some point the character would have to do something there, and would have to be something meaningful for the game – so that players did not simply run around climb over the House of Wisdom without being granted some insight into its historical and cultural significance, because it is too important.

Significant decisions like that were taken at the beginning, incorporated into pillars, and then, as Raphaël said, this can have an impact on all the different teams. There are also smaller decisions. But at some point, I would say that for me there are no really small decisions; as in the end I had to read each line of dialogue again and again.

Another example is with the voices in the game, as they are also incorporated as one of the pillars for both the audio team and the narrative team – the voices and the accents, the casting staff. I don't know if you want to dig into that right now, but I could talk about it for hours.

James: *I would like to ask you about the voices, and perhaps we could also take the chance to discuss the linguistic elements. For example, the game can be played entirely in Arabic with English subtitles, but even in the English audio version players can hear certain Arabic terms. For example, 'Caliph' is almost always referred to as 'Khalifa' in the English version, which might be unfamiliar to non-Arabic-speaking players, even though it is the correct formulation of the word in Arabic. How would the decision to use a term for caliph that Anglophone audiences may not understand be made? Essentially, who decides which terms to use and why?*

Sarah: It is a decision we made together, the creative director, the audio director, and myself. Just so you understand, the creative director oversees the whole core team. He leads the core team, like the director of a movie. And then you have all the directors for each department, as mentioned earlier. In the case of language, it was a decision we had to make together because it impacted both the writing team and the voice team.

In the video games industry, we have what we call the 'master language' – that is, the language in which lines are going to be written and voices recorded first. And then there is localisation – meaning that the master language may be translated to several different languages. In our case we wrote and record in English first.

The reason why we used language as we did in *Mirage* is because we wanted it to be as immersive as possible, since having an immersive city was one of our design pillars. With that in mind, we made three important choices. We first decided that every time we had the opportunity to put some Arabic words in the English audio version, we would do it, while making use of parenthetical translations[2] for a few times when this word was mentioned. We expected that players would eventually be able to understand some words just from hearing them, if a subtitled translation was provided a few times.

Another important point was that it could be tricky for anglophone speakers to correctly pronounce certain Arabic words. Therefore, the second decision was to cast English-speaking actors who were also native Arabic, Turkish, or Chinese speakers, depending on the characters – as we have some Turkish and Chinese characters. We wanted actors with authentic accents, and the authenticity team

---

2   The term parenthetical translation is used when a word or phrase is followed by its translation in another language in a pair of parentheses.

helped during casting sessions to ensure that the actors could speak Arabic fluently to the level of a native speaker.

The third decision was to not translate everything. For example, if you roam through the city of Baghdad in the game, you'll hear what we call 'exotic' languages. This means that if you pass through the House of Wisdom, you will hear people speaking Arabic, but also Chinese, Persian, and other languages, none of which are translated. This was done to make the experience as authentic and immersive as possible. So, while players may not understand everything, that is intentional and adds to the immersion. Interestingly, many native English speakers prefer playing in Arabic and reading the subtitles to experience the vibe of the city. Again, this decision was made very early on in the creative, narrative, and audio levels.

Raphaël: I would like to add a small thing about that, and the implications for the other teams. There is also the fact that when you change the language, frequently the same word takes longer to pronounce, which can affect the animation or timing of a sentence. For example, when translating from German to French to English, or from Arabic to English, the length of the sentence may vary. In some cases, certain words could not be included because they would either be too long or too short, which would disrupt the flow of the game.

Eduardo: I would like to end with a question on your views about games and learning, and my question is to both of you. As practitioners involved in making games that deal with the past, you have mentioned that a game has to be entertaining for players to deeply engage with it. From your perspective, at what particular moment is a game more suitable to turn into a tool for learning about historical aspects? Additionally, are there any particular aspects of historical knowledge or learning about history that, in your view as game producers, cannot be properly represented through a medium like a digital game?

Raphaël: I would say there are some ways in which video games are uniquely suited to present some forms of history to students or to whoever. When I was younger, I was attracted into the world of history by, not a game, but a movie called *Gladiator* (2000), which is set in Ancient Rome. Everybody knows that Gladiator is a really bad movie. At least from a purely historical perspective, it simply makes no sense. The Roman society presented there makes no sense. But when I was 12 and I watched that movie, it sparked an interest in the subject. From there, I went on to read books and to study at the university, and that was when I discovered that everything about it was just bad. But that was the moment: a big show that showed something and got me into history. Games can be this shiny, beautiful thing that you do for fun, and bring you into that world for a moment, and make you ask some questions about which you may decide to learn and have better answers later.

There are also some other things that video games can do, that movies, music, or books cannot. Video games offer a visual representation and a representation in which you are an actor. You interact with that world. You have to make choices and feel emotions while doing it, and all of that is a way to make you remember and creates a stronger link with what is being represented than simply reading it.

I feel some emotion when reading about a battle, but I have been a university professor in history. I am not a regular reader in that sense, I do think that people in 9th-century Baghdad are important. For me they are, but a game can reach someone else and impact them in a completely different way.

The choice-making part, the emotional part, is an important element in what videogames are good at. Then video games are bad at some other things. Regarding history, it is not a medium in which you can easily represent uncertainties. Especially in a game like *Assassin's Creed*, in which you are in an open world environment, but you do not completely control the narrative that the character will follow, and it is difficult to represent multiple interpretations.

Players can decide to go from point B to point A to point B, but the very pathways are not controllable. The story cannot completely tell the nuances the way you would want it to, and they are hard to convey. Therefore, nuances, uncertainties and, in general, the idea of a construction of discourse – that is what one can understand about the past – are difficult to communicate in a controlled way.

These are some things that video games are not that good at, but at the same time there are things that they are uniquely well suited to do.

Sarah: I would like to add a couple of things. First, I would argue that not every video game is meant to be fun. I really do not think so. *Assassin's Creed* is an example of a game that should be fun because it is for players who are aiming to have fun when they purchase the game. But there are several examples of games that are not at all fun. Apart from that point, I would not build a narrative solely around the educational aspect because I strongly believe that when a player roams a city they are unfamiliar with, and embarks on a long quest such as inside the House of Wisdom (to use that example again), they learn – almost without noticing – that this place was a key point of interest in Baghdad, where people translated books from different languages into Arabic, and so on.

Players learn without even noticing it. It is not like a lesson, it is more like, "Let's play with our PlayStation and, along the way, we are learning something". On a cognitive level, individuals learn better when they act on something, information is more effectively retained when we are involved in doing tasks. For example, as Raphael mentioned, every time a player makes a decision in the game – and exploration itself is a form of decision-making – whether they decide to explore a location or take the time to wander around, they will learn better than if they were simply watching a 20-minute scene inside the House of Wisdom or quickly skimming through a book, which they might not remember later. Each medium has a different implication.

There are things which are central to video games. I remember talking to some 15-year-olds who spoke English much better than I did at their age. I am

French, and generally, we are not great at learning languages. However, they spoke excellent English, often because they had been playing video games in English, sometimes without subtitles because they had no choice. They learned by playing, and that is also a natural way of learning.

Therefore, video games offer an important way of learning without players even realising it, and that is what we aim for in *Assassin's Creed*.

Biographical Notes

Jackson W. Armstrong is a professor of History at the University of Aberdeen, with a particular interest in late medieval Scotland and England.

James Baillie is a researcher at the Institute for Iranian Studies at the Austrian Academy of Sciences. His work covers the sociopolitical history of medieval Caucasia, development work on digital research methods, and research on how digital systems including games affect our understanding of the past.

Sarah Beaulieu was the narrative director for *Assassin's Creed Mirage*. Sarah is a writer for multiple media, and besides writing for games such as Assassin's Creed, she has written scripts for virtual reality and several other formats.

Linda Ryan Bengtsson is an associate professor in Media and Communication, member of the research centre for Geomedia Studies, Karlstad University.

Samantha Blackmon is an associate professor at Purdue University. Her intersectional work on games as cultural critique can be found in Feminist Media Histories, CCC, Computers and Composition, JAC, and CEA Critic. She is the co-founder of the award-winning Not Your Mama's Gamer, a five time Microsoft MVP, and winner of SIGDOC's Rigo Lifetime Achievement Award in the field of Communication Design for Game Design.

victoria l. braegger is an assistant professor of Technical Communication at Missouri University of Science and Technology. Her research focuses on usability, identity, and design in games, and highlights the impact of peripheral design on women and marginalised communities.

Eirik C. Brazier is associate professor of History at the University of South-Eastern Norway. His work explores the First World War and historical storytelling in video games, among other topics.

Osvaldo Cleger is associate professor of Spanish at the Georgia Institute of Technology. His research focuses on media history, digital culture, and the technopolitics of visuality in the Hispanic world.

Michael A. Conrad is a postdoctoral researcher at the University of Konstanz, affiliated with the Department of Literature, Art and Media Studies. While his current project focusses on cooperative gaming and other issues related to game studies, he has also published on medieval Iberia and transculturality.

William Hepburn is a lecturer in Medieval and Early Modern Scottish History at the University of Aberdeen.

Robert Houghton is a Senior Lecturer in Medieval History at the University of Winchester. His research concerns the representation of the Middle Ages and games and the utility of games as pedagogical, outreach and research tools.

Ragnhild Hutchison has since 2018 been CEO of Tidvis, a small Norwegian company that develops digital dissemination experiences focused on history, such as digital games and 3D visualisations of past places. She holds a PhD in History from the European University Institute and has worked extensively on economic and material history in the early modern period.

Eva Kingsepp is an associate professor in Media and Communication, independent scholar, most recently active at Stockholm University. Her main main research area concerns the relationship between history, memory and media, with a focus on the Second World War in popular culture, in particular digital games and films.

Diego Barroso Sánchez is a doctoral candidate studying the design of referential spaces in video games at the School of Creative Media, City University of Hong Kong. He is also interested in exploring how specific forms of agency express different ways of coupling reality and fiction.

Magnus H. Sandberg is an associate professor at the University of South-Eastern Norway, affiliated with the Department of Educational Science. His research examines the educational and cultural implications of digital media, particularly regarding learning, historical consciousness, and personal development.

Raphaël Weyland is a historian at Ubisoft Montréal. He worked as the lead historian advising on *Assassin's Creed Mirage*. Raphaël holds a PhD from the University of Montréal, with a thesis about the city of Seleucia on the Tigris during the Seleucid and Arsacid eras.

Acknowledgements

We would like to thank the reviewers of the articles of this issue of the Journal Digital Culture and Society for their diligent work and constructive feedback to the editors and authors.

Andrew Bearg, University of Houston Victoria
Jacob Bloomfeld, Universität Konstanz
Michael A. Conrad, Universität Konstanz
Iain Donald, Edinburgh Napier University
Mathias Fuchs, Leuphana Universität Lüneburg
Patrícia Gouveia, Universidade de Lisboa
Abbie Hartman, Macquarie University
Juan Hiriart, University of Salford
Daniel Kline, University of Alaska Anchorage
Victoria Lagrange, Kennesaw State University
Galen J. Lamphere-Englund, Extremism and Gaming Research Network
Jeffrey Lawler, California State University Long Beach
Eduardo Luersen, Universität Basel
Darren Reid, McGill University
Felipe Augusto Ribeiro, Universidade Federal de Pernambuco
Linda Schlegel, Peace Research Institute Frankfurt
Stefan Schubert, Universität Leipzig
Regina Seiwald, University of Birmingham
Alyssa Goldstein Sepinwall, California State University San Marcos
Sarah Slinguff, Walters Arts Museum
Micael Sousa, Universidade de Coimbra
Eve Stirling, Sheffield Hallam University
James Wilson, Rijksuniversiteit Groningen
K. T. Wong, Cornell University